# JOHN DEERE

## SHOP MANUAL

## More information available at haynes.com
### Phone: 805-498-6703

J H Haynes & Co. Ltd.
Haynes North America, Inc.

ISBN-10: 0-87288-567-4 ISBN-13:
978-0-87288-567-7

# Information and Instructions

This individual Shop Manual is one unit of a series on agricultural wheel type tractors. Contained in it are the necessary specifications and the brief but terse procedural data needed by a mechanic when repairing a tractor on which he has had no previous actual experience.

The material is arranged in a systematic order beginning with an index which is followed immediately by a Table of Condensed Service Specifications. These specifications include dimensions, fits, cleararances, capacities and tune-up information. Next in order of arrangement is the procedures section.

In the procedures section, the order of presentation starts with the front axle system and steering and proceeds toward the rear axle. The last portion of the procedures section is devoted to the power take-off and power lift systems.

Interspersed where needed in this section are additional tabular specifications pertaining to wear limits, torquing, etc.

## How to use the index

Suppose you wnt to know the procedure for R&R (remove and reinstall) of the engine camshaft. Your first step is to look in the index under the main heading of "Engine" until you find the entry "Camshaft." Now read to the right. Under the column covering the tractor you are repairing, you will find a number which indicates the beginning paragraph pertaining to the camshaft. To locate this paragraph in the manual, turn the pages until the running index appearing on the top outside corner of each page contains the number you are seeking. In this paragraph you will find the information concerning the removal of the camshaft.

# Common spark plug conditions

## NORMAL
**Symptoms:** Brown to grayish-tan color and slight electrode wear. Correct heat range for engine and operating conditions.
**Recommendation:** When new spark plugs are installed, replace with plugs of the same heat range.

### WORN
**Symptoms:** Rounded electrodes with a small amount of deposits on the firing end. Normal color. Causes hard starting in damp or cold weather and poor fuel economy.
**Recommendation:** Plugs have been left in the engine too long. Replace with new plugs of the same heat range. Follow the recommended maintenance schedule.

### CARBON DEPOSITS
**Symptoms:** Dry sooty deposits indicate a rich mixture or weak ignition. Causes misfiring, hard starting and hesitation.
**Recommendation:** Make sure the plug has the correct heat range. Check for a clogged air filter or problem in the fuel system or engine management system. Also check for ignition system problems.

### ASH DEPOSITS
**Symptoms:** Light brown deposits encrusted on the side or center electrodes or both. Derived from oil and/or fuel additives. Excessive amounts may mask the spark, causing misfiring and hesitation during acceleration.
**Recommendation:** If excessive deposits accumulate over a short time or low mileage, install new valve guide seals to prevent seepage of oil into the combustion chambers. Also try changing gasoline brands.

### OIL DEPOSITS
**Symptoms:** Oily coating caused by poor oil control. Oil is leaking past worn valve guides or piston rings into the combustion chamber. Causes hard starting, misfiring and hesitation.
**Recommendation:** Correct the mechanical condition with necessary repairs and install new plugs.

### GAP BRIDGING
**Symptoms:** Combustion deposits lodge between the electrodes. Heavy deposits accumulate and bridge the electrode gap. The plug ceases to fire, resulting in a dead cylinder.
**Recommendation:** Locate the faulty plug and remove the deposits from between the electrodes.

### TOO HOT
**Symptoms:** Blistered, white insulator, eroded electrode and absence of deposits. Results in shortened plug life.
**Recommendation:** Check for the correct plug heat range, over-advanced ignition timing, lean fuel mixture, intake manifold vacuum leaks, sticking valves and insufficient engine cooling.

### PREIGNITION
**Symptoms:** Melted electrodes. Insulators are white, but may be dirty due to misfiring or flying debris in the combustion chamber. Can lead to engine damage.
**Recommendation:** Check for the correct plug heat range, over-advanced ignition timing, lean fuel mixture, insufficient engine cooling and lack of lubrication.

### HIGH SPEED GLAZING
**Symptoms:** Insulator has yellowish, glazed appearance. Indicates that combustion chamber temperatures have risen suddenly during hard acceleration. Normal deposits melt to form a conductive coating. Causes misfiring at high speeds.
**Recommendation:** Install new plugs. Consider using a colder plug if driving habits warrant.

### DETONATION
**Symptoms:** Insulators may be cracked or chipped. Improper gap setting techniques can also result in a fractured insulator tip. Can lead to piston damage.
**Recommendation:** Make sure the fuel anti-knock values meet engine requirements. Use care when setting the gaps on new plugs. Avoid lugging the engine.

### MECHANICAL DAMAGE
**Symptoms:** May be caused by a foreign object in the combustion chamber or the piston striking an incorrect reach (too long) plug. Causes a dead cylinder and could result in piston damage.
**Recommendation:** Repair the mechanical damage. Remove the foreign object from the engine and/or install the correct reach plug.

# SHOP MANUAL
# JOHN DEERE

## MODELS 655, 755, 756, 855, 856 and 955

The 13 digit product identification (serial) number is on a plate (Fig.1) located below the rear pto shaft. The engine serial number plate (Fig. 2) is attached to top of rocker arm cover.

Fig. 1—The tractor serial number plate is located just below rear pto output shaft.

Fig. 2—Engine model and serial numbers are located on plate attached to top of rocker arm cover.

# INDEX (By Starting Paragraph)

# DUAL DIMENSIONS

This service manual provides specifications in both U.S. Customary and Metric (SI) systems of measurement. The first specification is given in the measuring system perceived by us to be the preferred system when servicing a particular component, while the second specification (given in parenthesis) is the converted measurement. For instance, a specification of 0.011 inch (0.28 mm) would indicate that we feel the preferred measurement in this instance is the U.S. Customary system of measurement and the Metric equivalent of 0.011 inch is 0.28 mm.

# CONDENSED SERVICE DATA

| | Models | | | |
|---|---|---|---|---|
| | 655 | 755<br>756 | 855<br>856 | 955 |

## GENERAL

| | | | | |
|---|---|---|---|---|
| Engine Make . . . . . . . . . . | | ─────────── Yanmar ─────────── | | |
| Engine Model . . . . . . . . . . | 3TN66UJ | 3TNA72UJ | 3TN75RJ | 3TN84RJ |
| Number of Cylinders . . . . . | | ─────── 3 ─────── | | |
| Bore . . . . . . . . . . . . . . . | 66 mm<br>(2.59 in.) | 72 mm<br>(2.83 in.) | 75 mm<br>(2.95 in.) | 84 mm<br>(3.31 in.) |
| Stroke . . . . . . . . . . . . . . | 64.2 mm<br>(2.53 in.) | 72 mm<br>(2.83 in.) | 75 mm<br>(2.95 in.) | 86 mm<br>(3.39 in.) |
| Displacement . . . . . . . . . . | 658 cc<br>(40.2 cu. in.) | 879 cc<br>(53.6 cu. in.) | 955 cc<br>(60.7 cu. in.) | 1430 cc<br>(87.3 cu. in.) |
| Compression Ratio . . . . . . | 22.4:1 | 22.3:1 | 17.8:1 | 18.0:1 |

## TUNE-UP

| | | | | |
|---|---|---|---|---|
| Firing Order* . . . . . . . . . . | | ─────── 1-3-2 ─────── | | |
| Valve Clearance, Cold —<br>  Inlet . . . . . . . . . . . . . . | | ─────── 0.2 mm ───────<br>(0.008 in.) | | |
| Exhaust . . . . . . . . . . . | | ─────── 0.2 mm ───────<br>(0.008 in.) | | |
| Valve Face and Seat Angle —<br>  Inlet . . . . . . . . . . . . . . | | ─────── 30° ─────── | | |
| Exhaust . . . . . . . . . . . | | ─────── 45° ─────── | | |
| Compression Pressure —<br>  Minimum . . . . . . . . . . | 2935 kPa<br>(426 psi) | | 2645 kPa<br>(384 psi) | |
| Injector —<br>  Opening Pressure . . . . . . | 11,242-12,202 kPa<br>(1630-1770 psi) | | 19,120-20,080 kPa<br>(2773-2913 psi) | |
| Engine Low Idle . . . . . . . . | | ─────── 1400-1500 rpm ─────── | | |
| Engine High Idle . . . . . . . . | | ─────── 3400-3450 rpm ─────── | | |
| Engine Rated Speed . . . . . | | ─────── 3200 rpm ─────── | | |
| Battery —<br>  Voltage . . . . . . . . . . . . . | | ─────── 12 ─────── | | |
| Terminal Grounded . . . . | | ─────── Negative ─────── | | |
| Transmission . . . . . . . . . . | | ── Hydrostatic and 2-speed range ── | | |
| Speeds . . . . . . . . . . . . . . | | ─────── Variable × 2 ─────── | | |

* Cylinders are numbered from back to front. Number 1 cylinder is rear cylinder.

## SIZES

| | | | | |
|---|---|---|---|---|
| Crankshaft Main<br>  Journal Diameter . . . . . . | 39.97-39.98 mm<br>(1.5736-1.5740 in.) | 43.97-43.98 mm<br>(1.731-1.732 in.) | 46.952-46.962 mm<br>(1.8485-1.8489 in.) | 49.952-49.962 mm<br>(1.9666-1.9670 in.) |
| Crankshaft Crankpin<br>  Diameter . . . . . . . . . . . . | 35.97-35.98 mm<br>(1.416-1.417 in.) | 39.97-39.98 mm<br>(1.5736-1.574 in.) | 42.952-42.962 mm<br>(1.691-1.6914 in.) | 47.952-47.962 mm<br>(1.8879-1.8883 in.) |
| Piston Pin Diameter . . . . . | 19.991-20.0 mm<br>(0.787-0.7874 in.) | 20.991-21.0 mm<br>(0.826-0.827 in.) | 22.991-23.0 mm<br>(0.905-0.906 in.) | 25.987-26.0 mm<br>(1.023-1.024 in.) |
| Valve Stem Diameter —<br>  Inlet . . . . . . . . . . . . . . | 5.460-5.475 mm<br>(0.215-0.216 in.) | 6.945-6.960 mm<br>(0.273-0.274 in.) | | 7.960-7.975 mm<br>(0.313-0.314 in.) |
| Exhaust . . . . . . . . . . . | 5.445-5.460 mm<br>(0.214-0.215 in.) | 6.945-6.960 mm<br>(0.273-0.274 in.) | | 7.960-7.975 mm<br>(0.313-0.314 in.) |

# CONDENSED SERVICE DATA (CONT.)

**Models**

| | 655 | 755<br>756 | 855<br>856 | 955 |
|---|---|---|---|---|
| **CLEARANCES** | | | | |
| Main Bearing —<br>Diametral Clearance.... | ——————0.020-0.0072 mm ——————<br>(0.0008-0.0028 in.) | | ——————0.038-0.093 mm——————<br>(0.0015-0.0037 in.) | |
| Rod Bearing —<br>Diametral Clearance.... | ——————0.020-0.0072 mm ——————<br>(0.0008-0.0028 in.) | | ——————0.038-0.090 mm——————<br>(0.0015-0.0035 in.) | |
| Camshaft Bearing,<br>Diametral Clearance —<br>Front.............. | 0.040-0.125 mm<br>(0.0016-0.0049 in.) | 0.040-0.085 mm<br>(0.00016-0.0033 in.) | ——————0.040-0.130 mm——————<br>(0.0015-0.0050 in.) | |
| Crankshaft End Play —<br>Standard ............. | 0.095-0.266 mm<br>(0.004-0.011 in.) | ——————————0.09-0.271 mm——————————<br>(0.004-0.011 in.) | | |
| **CAPACITIES** | | | | |
| Cooling System ......... | ——————————3.8 L——————————<br>(4 qt.) | | | 4.5 L<br>(4.8 qt.) |
| Crankcase<br>With Filter.......... | 2.4 L<br>(2.5 qt.) | 3.2 L<br>(3.4 qt.) | 3.9 L<br>(4.1 qt.) | 4.2 L<br>(4.5 L) |
| Transmission .......... | ——————————17 L ——————————<br>(4.5 gal.) | | | |
| Front Drive Axle ........ | —————— 2.13 L ——————<br>(2.25 qt.) | | | 3.3 L<br>(3.5 qt.) |

# FRONT AXLE SYSTEM (TWO-WHEEL DRIVE)

## FRONT AXLE ASSEMBLY AND STEERING LINKAGE

### Two-Wheel-Drive Models

**1. WHEELS AND BEARINGS.** Front wheel bearings should be removed, cleaned, inspected and renewed or repacked every 500 hours of operation. To remove front wheel hub and bearings, raise and support the front axle extension, then unbolt and remove the tire and wheel assembly. Remove cap (1—Fig. 3), cotter pin, castellated nut (2), washer (3) and outer bearing cone (4). Slide the hub assembly from spindle axle shaft. Remove seal (9) and inner bearing cone (8). Drive bearing cups (5 and 7) from hub (6) if renewal is required. Pack wheel bearings liberally with a suitable wheel bearing grease. Reassemble by reversing disassembly procedure. Tighten castellated nut (2) until a slight drag is felt, back nut off ¼ turn or just enough to install cotter pin, then install cap (1). On 955 models, tighten front wheel retaining nuts to 120 N•m (88 ft.-lbs.) torque. Tighten front wheel retaining nuts to 79 N•m (58 ft.-lbs.) torque for all 655, 755, 756, 855 and 856 models.

**2. TIE ROD AND TOE-IN.** A single tie rod connects left and right steering arms of spindles (10 and 27—Fig. 3). Automotive type tie rod ends are not adjustable for wear and should be renewed if worn. Rod ends threaded into tie rod are used to adjust the distance between ends and establish front wheel toe-in. Recommended toe-in is 6 mm (¼ in.) and should be measured between wheel rims on centerline of axle, parallel to ground. Rotate wheels and re-measure to be sure that wheels are not bent, giving incorrect reading. Tighten rod end jam nut to 118 N•m (87 ft.-lbs.) torque after toe-in is correctly set.

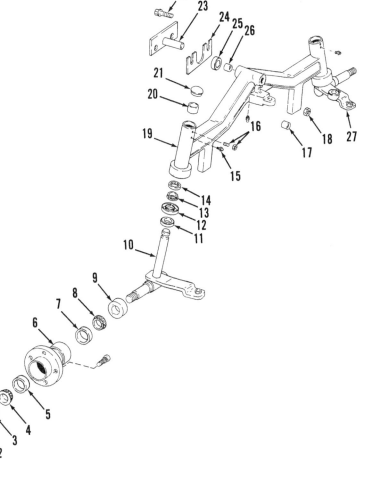

*Fig. 3—Exploded view of front axle typical of all two-wheel-drive models.*

1. Cap
2. Castellated nut
3. Washer
4. Outer bearing cone
5. Outer bearing cup
6. Hub
7. Inner bearing cup
8. Inner bearing cone
9. Seal
10. Left spindle
11. Thrust washer
12. Seal
13. Bearing
14. Bearing cup
15. Grease fitting
16. Set screw & locknut
17. Bushing (same as 26)
18. Locknut
19. Axle
20. Upper bushing
21. Cap
22. Screws
23. Pivot pin
24. Shim
25. Sleeve
26. Bushing (same as 17)
27. Right spindle

**3. SPINDLES AND BUSHINGS.** To remove spindle (10—Fig. 3), first remove wheel and hub. Disconnect rod end from steering arm and remove cap (21). Steering cylinder must be disconnected if right spindle (27) is removed. Loosen locknut, remove set screw (16), then remove spindle from axle. Clean and inspect parts for wear or other damage and renew as necessary.

Spindle upper bushing (20) should be pulled from top using a 31.75-38.1 mm (1¼-1½ in.) blind hole puller. Drive new bushing into bore until it bottoms against shoulder. Lower bearing race (14) can be pulled from axle bore if renewal is required. Drive new race into axle bore until it bottoms against shoulder, pack bearing (13) with grease, then install bearing and seal (12). Clean spindle, remove thrust washer (11) and install new thrust washer with outside groove toward top. Insert spindle into axle bore and install retaining set screw (16). Tighten set screw until tight, loosen set screw ⅛-¼ turn, then tighten locknut. Install cap (21), attach steering tie rod ends to steering arms, tighten castellated nuts to 53 N·m (39 ft.-lbs.) torque and install cotter pin. Attach steering cylinder to right steering arm, tighten castellated nut to 75 N·m (55 ft.-lbs.) torque and install cotter pin.

Refer to paragraph 2 for toe-in adjustment. Tighten front wheel retaining nuts of 955 models to 120 N·m (88 ft.-lbs.) torque. Tighten front wheel retaining nuts to 79 N·m (58 ft.-lbs.) torque for all 655, 755, 756, 855 and 856 models.

**4. AXLE MAIN MEMBER, PIVOT PIN AND BUSHINGS.** To remove front axle assembly, first remove any front mounted equipment and weights. Disconnect battery ground cable. Remove fuel tank from all 655, early 755, all 756, early 855 and all 856 models. Remove battery from late 755, late 855 and all 955 models. On all models, detach both ends of steering cylinder, but leave hoses attached. Raise and block front of tractor in such a way that it will not interfere with the removal of axle. Remove wheels and support axle in the center with a floor jack so that it can be lowered and moved away from tractor. Remove locknuts (18—Fig. 3) and screws (22), then carefully remove shims (24). Remove pivot pin (23), carefully lower the axle assembly and roll axle from under tractor.

Check axle pivot bushings (17 and 26) and renew if necessary. Bushings are pressed into bore of axle and should be installed flush with bore. Reverse removal procedure when assembling. Axle end play should be less than 1 mm (0.04 in.), but should have slight clearance. To measure and adjust end play, push the axle to the rear, then measure clearance between front of axle and plate on pivot pin with a feeler gauge. Shims (24) should be added to increase clearance. Make sure that screws (22) attaching plate (23) are tight when measuring, but that axle does not bind. Tighten screws (22) and locknuts (18) to 130 N·m (96 ft.-lbs.) torque. Refer to paragraphs 2 and 3 for additional torque values and assembly notes.

# STEERING SYSTEM

**5.** A hydrostatic steering system is used, which consists of a steering valve assembly and steering cylinder. Pressurized oil is supplied to the steering control valve by the hydraulic pump via proportional flow divider. In the event of engine stoppage, steering can be accomplished by means of the gerotor pump located in the steering valve.

Fluid and canister filter for the hydraulic system, steering system and hydrostatic transmission should be changed and transmission vent tube (1—Fig. 4) and screen (11) should be removed and cleaned after 500 hours of operation.

It may be necessary to bleed air from transmission, hydraulic and steering systems as follows after draining the system or installing a dry repair part. Fill the system, start engine and operate at high idle speed for one minute. Turn steering wheel full left and hold for five seconds, turn steering wheel until wheels

point straight forward and hold for ten seconds, then turn steering wheel to full right position and hold for five seconds. Return wheels to straight ahead position and move tractor forward about 6 meters (20 ft.), make two hard left turns, then make two hard right turns. Move tractor about ten feet in reverse, then shut off engine and inspect for oil leaks. Recheck oil level by and if necessary fill with "John Deere Low Viscosity HY-GARD" transmission and hydraulic oil or equivalent.

## STEERING CONTROL VALVE

### All Models

**6. REMOVE AND REINSTALL.** Power steering control valve is located at lower end of steering wheel shaft. To remove the steering control valve and steering shaft, proceed as follows. Remove muffler, pedestal side panels and steering wheel. Remove the battery and battery base from all 655, early 755, all 756, early 855 and all 856 models. Remove the fuel tank from late 755, late 855 and all 955 models. Remove the engine to transmission drive shaft from all models as outlined in paragraph 24. Relieve oil pressure from hydraulic system by lowering any mounted equipment and cycling hydraulic control valves. Mark the four hydraulic lines at bottom of steering control valve to facilitate correct reattachment, then detach lines. Cover all openings to prevent the entrance of dirt. Remove nuts securing steering valve to support and withdraw unit.

Reinstall steering valve and column by reversing removal procedure. Plug for relief valve should be toward right side of tractor. Install steering wheel and tighten retaining nut to 13-16 N·m (10-12 ft.-lbs.) torque. Use new "O" rings to seal hydraulic line connections. Refer to paragraph 24 and reinstall drive shaft.

**7. OVERHAUL.** To disassemble, the removed steering control valve should be attached to a holding fixture, which can be clamped in a vise. **Steering control valve will be distorted and damaged if clamped directly in vise.** Refer to Fig. 5 for drawing of fixture which can be locally fabricated and used to hold steering control valve while disassembling and reassembling.

Before disassembling steering valve, note alignment grooves machined in side of unit (Fig. 6). Attach steering control valve to holding fixture with four nuts. Clamp holding fixture in vise with steering shaft down. Remove the four nuts securing port cover (1—Fig. 7), then lift cover from studs. Remove seal ring (5) and "O" rings (6). Remove plug (4) and relief valve (2). Remove "O" ring (3) from plug. Remove port manifold (7) and the three springs (9). Springs (9) are

*Fig. 4—Partially exploded view of transaxle center housing showing some of the component parts.*

1. Vent
2. Fill plug
3. Rockshaft housing
4. Side drain plug
5. Dipstick
6. Hydraulic pump
7. Rear drain plug
8. Adapter
9. Back-up ring
10. "O" ring
11. Filter screen
12. Transaxle housing rear cover
13. Tranaxle housing

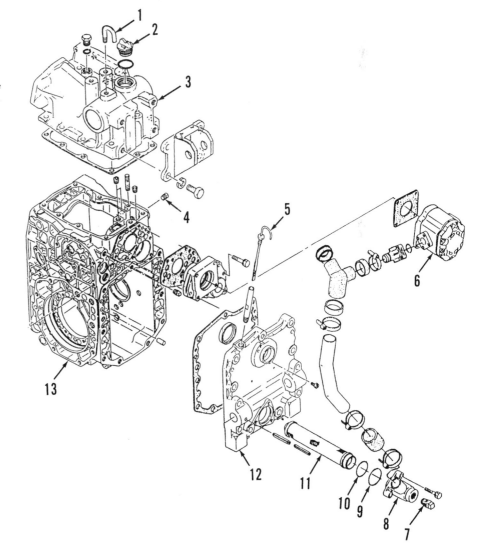

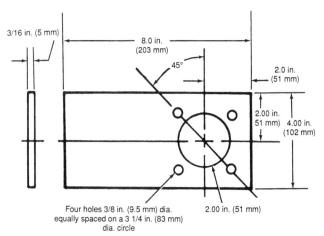

*Fig. 5—Fabricate a holding fixture as shown to hold steering control valve while disassembling and reassembling. Valve can be damaged by clamping directly in vise.*

19 mm (¾ in.) long and must only be replaced as a complete set with similar, but shorter, springs (14). Inspect machined surfaces of manifold (7) for scratches or grooves. All edges of manifold must be sharp and free of nicks or burrs.

Remove alignment pins (8) and valve ring (11). New seal rings (10 and 12) should be installed when reas-

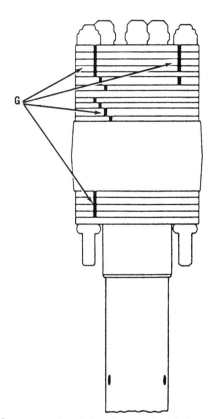

*Fig. 6—Components of steering control valve which have alignment grooves (G) must be installed with grooves positioned as shown. Unit will not operate properly if parts are improperly aligned.*

sembling. Remove valve plate (13) and inspect surfaces for scratches or grooves. All edges of plate must be sharp. Remove hex drive assembly (16) and pin (15). Check hex drive assembly for wear, grooves or scoring. Pin should be tight and show no wear or damage. Remove the three springs (14) and check for broken coils, wear or other damage. Springs (14) are 13 mm (½ in.) long and must only be replaced as a complete set with similar, but longer, springs (9).

Remove isolation manifold (17) and inspect machined surfaces for scratches or grooves. Isolation manifold is four plates bonded together. **Do not attempt to separate the four plates.** All machined edges of isolation manifold must be sharp and free of nicks or burrs.

Remove drive link (18) and check crowned surfaces for wear or scoring. Lift metering ring (20) from studs. New seals (19 and 37) should be installed when assembling. Install new metering ring if machined bore is scored. Carefully lift metering assembly (23 through 32) from steering column and other parts attached to holding fixture, then set metering assembly on clean soft surface.

**NOTE: Do not clamp metering assembly in vise. Parts of the metering assembly can be easily damaged by improper handling.**

Remove seal (24) and Allen head screws (23), then lift commutator cover (25) off. Check machined surfaces of commutator cover for nicks, burrs, scoring other excessive wear. Surface should be polished because of rotation.

Carefully lift commutator ring (26) from stator and check for wear, nicks, burrs or cracks. Commutator ring is fragile and can be easily broken. Carefully remove commutator (27) from the five pins (28) using a small wood stick or equivalent.

**NOTE: The commutator is easily damaged by prying with a screwdriver or other hard object. The commutator is manufactured by bonding two plates together, but do not attempt to separate the two parts.**

Check commutator for any sign of wear. All surfaces should be smooth and all edges sharp. If either commutator (27) or commutator ring (26) is damaged, both must be renewed as a matched set.

Remove drive link spacer (29) and check for damage. Rotor (30) should move freely inside stator (31) and tip clearance should be less than 0.08 mm (0.003 in.). Check all surfaces for wear or scoring and install new set (30 and 31) if either is damaged.

All surfaces of drive plate (32) should be smooth and should show little more than a polished surface. Remove thrust bearing (33), bearing spacer (34), seal spacer (35) and seal (36). Machined surface of upper cover plate (38) should be smooth and free from nicks and burrs.

Clean and inspect all parts. If wear is general, it is often better to replace the complete assembly than individual components. When reassembling, renew all seals and "O" rings and make sure all parts are lubricated and clean.

Begin assembly by attaching the four studs (43—Fig. 7) to holding fixture if they were removed. Retaining nuts need to be only finger tight. Bushing at top end of steering column should be 2.5 mm (0.1 in.) below edge of column. Bend edge of column tube in slightly to retain bushing after installation. Lubri-cate inside of bushing with multipurpose grease. Slide steering column over studs and make sure that square holes in column seat firmly on square shoulder of studs. Coat washer (40) with grease and position in recess of column. Install snap ring (39) in groove of steering shaft and insert shaft into column.

Install upper cover plate (38) with highly polished surface up. Make note of the location of alignment grooves (G—Fig. 6) in plate (38—Fig. 7). Grease the polished surface of upper plate (38), then position thrust bearing (33) and spacer (34). Install seal spacer (35), face seal and seal back-up ring (36). Refer to Fig. 8 for cross-sectional view of seal assembly.

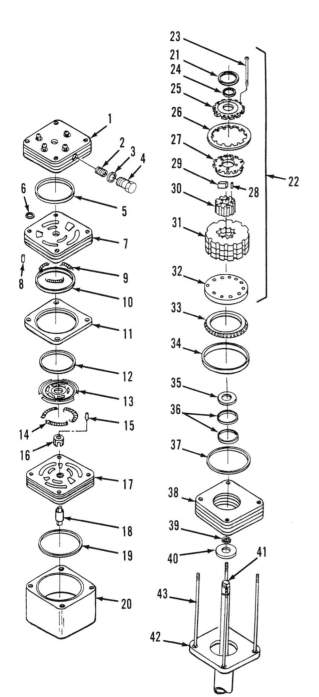

**Fig. 7—Exploded view of hydrostatic steering control valve. The three springs (9) are 19 mm (3/4 in.) long and the three springs (14) are 13 mm (1/2 in.) long.**

| | |
|---|---|
| 1. Port cover | 23. Screws (11 used) |
| 2. Relief valve | 24. Commutator seal |
| 3. "O" ring | 25. Commutator cover |
| 4. Plug | 26. Commutator ring |
| 5. Seal ring | 27. Commutator |
| 6. "O" rings | 28. Alignment pins (5 used) |
| 7. Port manifold | 29. Drive link spacer |
| 8. Alignment pins (2) | 30. Rotor |
| 9. Springs—19 mm (3/4 in.) | 31. Stator |
| 10. Seal ring | 32. Drive plate |
| 11. Valve ring | 33. Thrust bearing |
| 12. Seal ring | 34. Thrust bearing spacer |
| 13. Valve plate | 35. Seal spacer |
| 14. Springs—13 mm (1/2 in.) | 36. Face seal & back-up ring |
| 15. Pin | 37. Seal |
| 16. Hex drive assy. | 38. Upper cover plate |
| 17. Isolation manifold | 39. Snap ring |
| 18. Drive link | 40. Washer |
| 19. Seal | 41. Steering shaft |
| 20. Metering ring | 42. Steering tube assy. |
| 21. Seal | 43. Holding studs |
| 22. Metering assy. | |

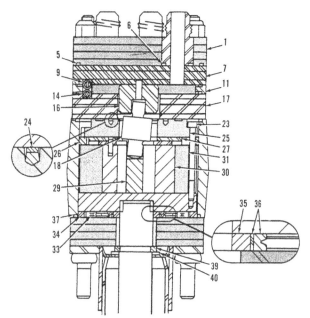

Fig. 8—Cross section of steering control valve. Refer to Fig. 7 for legend.

bly (22) into ring (20), then turn complete assembly over. Cut 0.18 mm (0.007 in.) thick shim stock into six pieces, each approximately $13 \times 38$ mm ($\frac{1}{2} \times 1\frac{1}{2}$ in.). Install shims (S—Fig. 9) at three positions 120° apart between inside bore of metering ring (20) and the stator and drive plate (32). Two shims should be at each of the three locations. After shims are installed, and parts are centered, turn assembly over and tighten the 11 Allen screws evenly in several small steps until final torque of 1.3-1.5 N•m (11-13 in.-lbs.) is reached. Tighten screws in sequence shown in Fig. 10, then remove shims used to center the metering assembly. Remove metering assembly from metering ring. Insert drive link (18—Fig. 7) into rotor (30), hold stator, then turn the drive link and rotor by hand. The rotor should turn freely and easily with no binding. If any binding is encountered, disassemble the metering assembly, determine the cause, correct the problem, reassemble and recheck. Improper or uneven tightening of the 11 Allen screws can cause binding. Refer to Fig. 8 for cross-sectional view.

Coat seal ring (37—Fig. 7) with multipurpose grease and seat the seal ring in groove in end of metering ring (20), which has no pin holes. Install metering ring (20) over studs (43) with one pin hole toward top and aligned with groove (G—Fig. 6) in upper cover plate (38—Fig. 7). Grease surface of drive plate (32) and lower into metering ring (20). Turn metering assembly until drive plate engages steering shaft. When properly installed, surface of metering assembly will be below surface of metering ring. Coat commutator seal ring (24) with grease and install in cover (25) with yellow marked part of seal down, into cover. Grease seal ring (19), then seat in groove of metering ring. Install the two alignment pins in holes of metering ring and install isolation manifold (17).

Place drive plate (32—Fig. 7) on clean dry surface with slotted side down. Position stator (31) on drive plate so that slots are aligned with threaded holes in plate. Position rotor (30), with five pin holes up, inside stator, then grease drive link spacer (29) and insert in rotor slot. Install commutator (27), with long slots up, align holes in commutator and rotor, then install the five pins (28). Install commutator ring (26) and commutator cover (25) with holes aligned with slots in stator. Flat surface of cover (25) should be toward commutator. Clean threads in drive plate with thread locking clean and prime solution, apply low strength thread locking compound to threads of the 11 Allen screws (23) and start all 11 screws. **Do not tighten screws yet.**

Align parts of the metering assembly as follows while tightening screws (23). Slide metering assem-

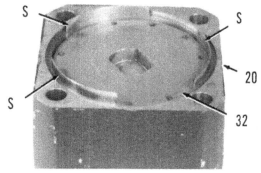

Fig. 9—Center parts of metering assembly using 0.36 mm (0.014 inch) thickness of shims at three locations as shown at (S).

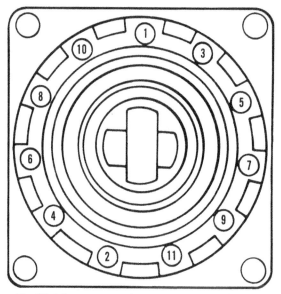

Fig. 10—Tighten screws in several steps in sequence shown to reduce distortion.

Recessed slots in isolation manifold should be up and grooves (G—Fig. 6) should be aligned as shown. Install the two alignment pins in holes of isolation manifold and the three 13 mm (½ in.) long springs (14—Fig. 7) in recessed slots. Install drive hex (16) over the previously installed drive link (18). Make sure pin (15) is installed and up. Coat seal rings (10 and 12) with multipurpose grease and install on valve ring (11). Install valve ring with seal rings over the two alignment pins and with grooves (G—Fig. 6) aligned as shown. Identify side of valve plate (13—Fig. 7) that is marked "PORT SIDE." Install valve plate with marked side up, toward port manifold (7) and 180° (opposite) the side marked with grooves (G—Fig. 6). Move the valve plate to be sure that plate is centered over springs (14—Fig. 7). Install the three 19 mm (¾ in.) long springs (9) in recessed slots of port manifold (7), then install manifold with grooves (G—Fig. 6) aligned as shown. Be careful to make sure alignment pins (8—Fig. 7) and pin (15) correctly enter holes in port manifold (7). Coat "O" rings (6) and seal ring (5) with grease and position in grooves of port cover (1), then install port cover with grooves (G—Fig. 6) aligned.

Check all of the alignment grooves (G) to make sure that all parts are assembled correctly before proceeding. Install the four retaining nuts and tighten in diagonal pattern, gradually increasing torque until final torque of 27-33 N·m (239-293 in.-lbs.) is reached. Make sure that "O" rings and seals are not out of position while tightening. Install relief valve (2—Fig. 7) using new seal ring (3) on plug. If correctly assembled, steering shaft should rotate freely without binding.

## STEERING CYLINDER

### All Models

**8. R&R AND OVERHAUL.** Mark hydraulic hose locations and disconnect hoses. Disconnect steering cylinder at each end and remove cylinder. Steering cylinder used is a welded assembly and must be renewed as an entire unit.

# FRONT AXLE SYSTEM (FOUR-WHEEL DRIVE)

**9.** Mechanical front-wheel drive is available. There may be differences between the front-wheel-drive systems as noted in the servicing instructions which follow.

## MAINTENANCE

### Four-Wheel-Drive Models

**10.** Oil should be maintained at top mark of dipstick attached to fill plug (F—Fig. 11). Every 200 hours, drain oil from front drive axle by removing the two plugs (D) and plug (P) from housings. Fill with "John Deere GL-5" or equivalent gear lubricant to correct level. Capacity is approximately 2.5 L (2.6 qt.) for 955 models; 2.13 L (2.25 qt.) for other models; however, some lubricant may be trapped in housing.

## FRONT AXLE ASSEMBLY AND STEERING LINKAGE

### Four-Wheel-Drive Models

**11. TIE ROD AND TOE-IN.** A single tie rod (9—Fig. 11) connects left and right steering arms (4 and 10). Automotive type ends are not adjustable for wear and should be renewed if worn. Attach steering tie rod ends to steering arms, tighten castellated nuts to 53 N·m (39 ft.-lbs.) torque and install cotter pins.

Recommended toe-in is 3-9 mm (⅛-⅜ in.) and is adjusted by threading rod ends into tie rod to change the distance between ends. To check alignment, measure between wheel rims on centerline of axle, parallel to ground. Rotate wheels and re-measure to be sure that wheels are not bent, giving incorrect reading. Tighten rod end jam nut to 118 N·m (87 ft.-lbs.) torque after wheel alignment is correctly set.

**12. R&R AXLE ASSEMBLY.** To remove front axle assembly, first remove any front mounted equipment and weights. Disconnect battery ground cable. Remove fuel tank from all 655, early 755, all 756, early 855 and all 856 models. Remove battery from late 755, late 855 and all 955 models. On all models, detach both ends of steering cylinder, but leave hoses attached. Raise and block front of tractor in such a way that it will not interfere with removal of axle. Support tractor behind axle and remove both front wheels. If unit is to be disassembled, remove plugs (D—Fig. 11) from final drive housings and drain oil from front axle assembly. Remove the front axle drive shaft. Support axle level with floor to prevent tipping, in such a way that it can be lowered and moved away from tractor. Remove screws (22), then carefully remove shims (24). Remove pivot pin (23), carefully lower the axle assembly and roll axle from under tractor.

Check axle pivot bushings (26) and renew if necessary. Bushings are pressed into bore of housing (7) and should be installed flush with bore. Reverse removal procedure when assembling. Axle end play should be 0.127-1.016 mm (0.005-0.040 in.). To measure and adjust end play, push axle to the rear, then measure clearance between front of axle and plate on pivot pin with a feeler gauge. Shims (24) should be added to increase clearance. Make sure that screws (22) attaching plate (23) are tight when measuring, but that axle does not bind. Tighten screws (22) to 88 N·m (65 ft.-lbs.) torque. Attach steering cylinder to right steering arm, tighten castellated nuts to 75 N·m (55 ft.-lbs.) torque and install cotter pins. Tighten

front wheel retaining nuts of 955 models to 120 N·m (88 ft.-lbs.) torque. Tighten front wheel retaining nuts to 79 N·m (58 ft.-lbs.) torque for all 655, 755, 756, 855 and 856 models. Refer to paragraph 11 for additional torque values and assembly notes.

## FRONT FINAL DRIVE

### Four-Wheel-Drive Models

**13. R&R AND OVERHAUL.** Raise and block front of tractor in a way that is safe and will not interfere with the removal of front wheels or final drive. Re-

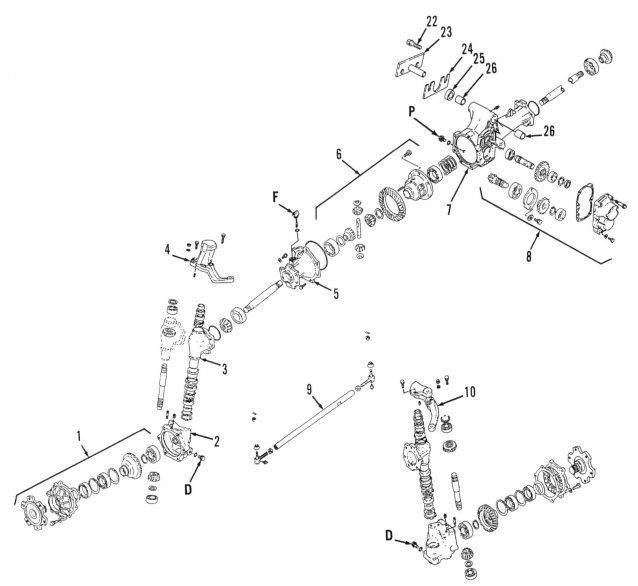

*Fig. 11—Exploded drawing of mechanical front wheel drive axle. Oil can be drained from axle by removing drain plugs (D & P). Dipstick is attached to fill plug (F) for measuring oil level.*

| | | | |
|---|---|---|---|
| 1. Final drive & wheel axle | 5. Left axle housing | 8. Differential input | 23. Pivot pin |
| 2. Spindle gear case | 6. Differential | 9. Tie rod | 24. Shim |
| 3. Spindle housing | 7. Differential & right | 10. Right steering arm | 25. Sleeve |
| 4. Left steering arm |    axle housing | 22. Screws | 26. Bushings |

move drain plug (D—Fig. 12) and permit oil to drain from axle housing. Remove the six screws that attach final drive housing (12) to spindle housing (2), then separate housings. Bearing (21) will probably remain in spindle housing, but can be removed with puller. Outer race of bearing (21) is located in bore of gear (19). Wheel axle and hub (11) can be pressed from gear (19) after removing snap ring (20). Wear sleeve (14) can be removed if renewal is required. Gear (19) can be pressed from bearing (17), then bearing (17) can be pressed from housing after removing seal (15) and snap ring (16). Be sure to remove old gasket (13) and install new gasket.

Press new bearing (17) against snap ring (18), then install snap ring (16). Coat lip of seal (15) with grease and press into housing bore. Press wheel axle and hub (11) into gear (19), then install snap ring (20). Install drain plug (D) and new gasket (13), then insert wheel axle and housing into spindle housing (2). Arrow cast into housing (12) should point **up** and be toward top. Tighten retaining screws in a crossing pattern to 52 N·m (38 ft.-lbs.) torque. Fill axle with "John Deere GL-5" or equivalent gear lubricant to correct level marked on dipstick of fill plug (F—Fig. 11). Capacity is approximately 2.5 L (2.6 qt.) for 955 models; 2.13 L (2.25 qt.) for other models; however, some lubricant may be trapped in housing. Tighten front wheel retaining nuts of 955 models to 120 N·m (88 ft.-lbs.) torque. Tighten front wheel retaining nuts to 79 N·m (58 ft.-lbs.) torque for all 655, 755, 756, 855 and 856 models.

## FRONT DRIVE SPINDLE

### Four-Wheel-Drive Models

**14. R&R AND OVERHAUL.** To remove the front drive spindle assembly, first refer to paragraph 13 and remove front final drive. Disconnect tie rod end from steering arm (4 or 10—Fig. 11). Steering cylinder must be disconnected if right spindle is removed. Remove six screws which attach spindle housing (3) to axle housing (5 or 7) and remove the complete spindle assembly. Drive shaft (48—Fig. 12) can be withdrawn if required.

*Fig. 12—Exploded view of mechanical front wheel drive left side spindle and final drive gear case. Parts for right side are similar. Washer (33), located between bearing (32) and snap ring (37), is used on all axles above serial number 017048 and all replacement gear cases.*

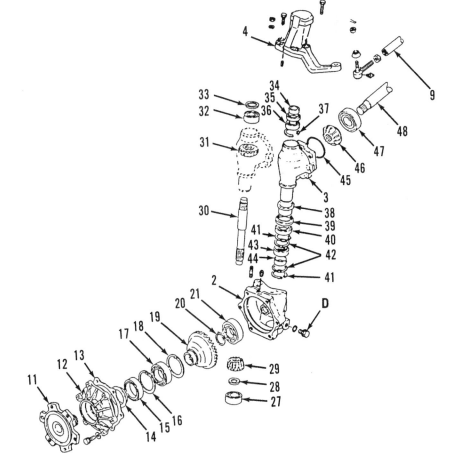

2. Spindle gear case
3. Spindle housing
4. Left steering arm
9. Tie rod
11. Wheel hub & axle
12. Final drive housing
13. Gasket
14. Sleeve
15. Seal
16. Snap ring
17. Ball bearing
18. Snap ring
19. Bevel gear
20. Snap ring
21. Ball bearing
27. Ball bearing
28. Washer
29. Final drive pinion
30. Spindle shaft
31. Driven gear
32. Ball bearing (same as 36)
33. Washer
34. Oil seal wear sleeve
35. Seal
36. Ball bearing (same as 32)
37. Snap ring
38. Oil seal wear sleeve
39. Bushing
40. Oil seal
41. Snap rings (2 used)
42. Washers (2 used)
43. Needle bearing
44. Inner race
45. Seal ring
46. Drive gear
47. Ball bearing
48. Left drive shaft

Remove the two nuts and two screws which attach steering arm (4 or 10—Fig. 11) to spindle gear case (2). Use a soft hammer to bump steering arm from dowel pins. Remove wear sleeve (34—Fig. 12), from shaft of steering arm only if new wear sleeve is to be installed. Seal (35) can be pulled from bore in spindle housing.

Bump spindle gear case (2) from spindle housing (3). Spindle shaft (30) may remain in spindle housing or in spindle gear case. Procedure for further disassembly will vary, but will be obvious. Refer to Fig. 12.

When assembling, press bearing (27) into bottom bore of gear case (2), then position washer (28) and gear (29) in case above the installed bearing. Gear (29) has 12 teeth for 955 models; 13 teeth for other models. Install snap ring (41) in lower groove, followed by lower washer (42). Press needle bearing (43) into bore against the washer, then install inner race (44). Install upper washer (42) above the installed needle bearing, then install upper snap ring (41) in groove. Coat lip of seal (40) with grease, then press seal into bore until it seats against snap ring. Press bushing (39) into bore against seal.

Carefully press oil seal wear sleeve (38) onto spindle housing (3) until fully seated. Press bearing (32) into bore until it is just below snap ring groove, then install washer (33) and snap ring (37).

**NOTE: Some early models may not have washer (33).**

Press bearing (36) into bore against snap ring, grease lip of seal (35), then press seal into bore against bearing. Oil seal wear sleeve (34) is pressed onto post of steering arm (4 or 10—Fig. 11).

Position upper (16 tooth) gear (31—Fig. 12) in spindle housing (3), then insert spindle shaft (30) through spindle housing with shorter splines in gear (31). Upper end of shaft should enter bearing (32). Tap shaft to seat shaft in bearing and gear.

Make sure that lower gear (29) and washer (28) are correctly positioned over the installed bearing (27), then insert spindle shaft and housing assembly into gear housing (2). Make sure that shaft splines are aligned with gear and end of shaft enters the installed lower bearing (27). Install steering arm (4 or 10—Fig. 11), but do not tighten retaining nuts and screws yet. Insert drive gear (46—Fig. 12) and bearing (47) into bore of spindle housing (3) and install new seal ring (45).

If removed, install drive shaft (48) with short splines out toward gear (46). Position spindle assembly and bearing (47) against axle housing, then install retaining screws. Tighten the six screws retaining spindle to axle to 52 N·m (38 ft.-lbs.) torque, then tighten the two nuts and two screws retaining steering arm to 52 N·m (38 ft.-lbs.) torque.

Install bearing (21), drain plug (D) and new gasket (13). Insert wheel axle and housing into spindle housing (2), making sure that arrow cast into housing (12) is pointing **up**, toward top. Tighten retaining screws in a crossing pattern to 52 N·m (38 ft.-lbs.) torque. Fill axle with "John Deere GL-5" or equivalent gear lubricant to correct level marked on dipstick of fill plug (F—Fig. 11). Capacity is approximately 2.5 L (2.6 qt.) for 955 models; 2.13 L (2.25 qt.) for other models; however, some lubricant may be trapped in housing. Tighten front wheel retaining nuts of 955 models to 120 N·m (88 ft.-lbs.) torque. Tighten front wheel retaining nuts to 79 N·m (58 ft.-lbs.) torque for all 655, 755, 756, 855 and 856 models.

## DIFFERENTIAL INPUT

### Four-Wheel-Drive Models

**15. R&R AND OVERHAUL.** Refer to paragraph 12 and remove the front drive axle. Remove drain plug (P—Fig. 13) and drain lubricant from axle center housing. If other disassembly is required, remaining lubricant should be drained by also removing final drive housing drain plugs (D—Fig. 11). Remove the five regular cap screws and two special piloted cap screws retaining cover (49—Fig. 13) to axle housing, then remove differential input cover. Shaft (55), gear (54) and bearings may be removed with cover, but if not, shaft and bearing (56) can be withdrawn after removing cover. Bearings (53, 56 and 57) can be

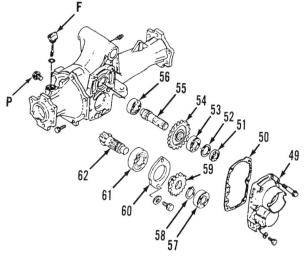

*Fig. 13—Exploded view of front axle differential input.*

| | | | |
|---|---|---|---|
| 49. | Drive housing cover | 56. | Ball bearing (same as 53) |
| 50. | Gasket | 57. | Ball bearing |
| 51. | Oil seal | 58. | Snap ring |
| 52. | Washer | 59. | Gear |
| 53. | Ball bearing (same as 56) | 60. | Retainer |
| 54. | Gear | 61. | Ball bearing |
| 55. | Shaft | 62. | Drive pinion & shaft |

removed from shafts using a suitable puller. Do not lose washer (52) which is located between bearing (53) and bore in cover. Gear (59) can be pulled from shaft after removing snap ring (58). Remove screws attaching retainer (60) to axle housing, then remove retainer, bearing (61) and bevel pinion shaft (62).

When reassembling, notch in retainer (60) must be correctly positioned to permit installation of cover (49). Screws attaching retainer (60) should be tightened to 26 N·m (19 ft.-lbs.) torque. Install new gasket (50) and attach cover (49) with the two piloting cap screws in the upper right and lower left positions. Install the five standard retaining screws in other locations. First, tighten the two piloting cap screws to 26 N·m (19 ft.-lbs.) torque, then tighten the five standard retaining screws to the same torque. Refer to paragraph 12 for installing axle assembly. Install drain plugs and fill axle with "John Deere GL-5" or equivalent gear lubricant to correct level marked on dipstick of fill plug (F—Fig. 11). Capacity is approximately 2.5 L (2.6 qt.) for 955 models; 2.13 L (2.25 qt.) for other models; however, some lubricant may be trapped in housing. Tighten front wheel retaining nuts of 955 models to 120 N·m (88 ft.-lbs.) torque. Tighten front wheel retaining nuts to 79 N·m (58 ft.-lbs.) torque for all 655, 755, 756, 855 and 856 models.

## DIFFERENTIAL

### Four-Wheel-Drive Models

**16. R&R AND OVERHAUL.** To remove the differential, first refer to paragraph 12 and remove axle assembly. Remove final drives and steering spindles as outlined in paragraphs 13 and 14. Withdraw both drive shafts (48 and 63—Fig. 14) from axle housings. Refer to paragraph 15 and remove differential input cover (49) and shafts (55 and 62).

Unbolt left axle housing (5) and separate from differential housing and right axle housing (7). Differential assembly can be lifted from housing. Be careful not to lose shims (75) which sets backlash between bevel pinion (62) and ring gear (71).

Use puller to remove bearings (65 and 74). Ring gear (71) is attached to differential case (73) with six screws. Drive roll pin (72) from case and cross shaft (70), then drive cross shaft from case and gears (68). Roll gears (68) and thrust washers (69) from differential case, then withdraw side gears (67) and thrust washers (66).

Assembly of differential is reverse of disassembly procedure. Apply medium strength thread lock to screws retaining ring gear (71) and tighten to 22 N·m (16 ft.-lbs.) torque.

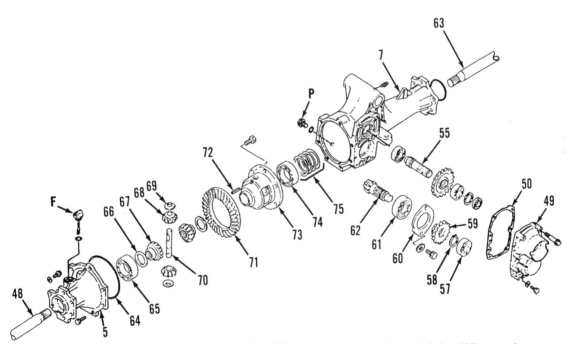

**Fig. 14—Exploded view of axle housings and differential. Differential input and bevel pinion (62) must be removed before differential can be removed from housing (7).**

| | | |
|---|---|---|
| 5. Left axle housing | 57. Bearing | 69. Thrust washer (2 used) |
| 7. Differential & right axle housing | 58. Snap ring | 70. Cross shaft |
| 48. Left drive shaft | 59. Gear | 71. Ring gear |
| 49. Drive housing cover | 60. Retainer | 72. Roll pin |
| 50. Gasket | 61. Bearing | 73. Differential |
| 55. Shaft | 62. Drive pinion & shaft | 74. Ball bearing (same as 65) |
| | 63. Right drive shaft | 75. Shims |
| | 64. Seal ring | |
| | 65. Ball bearing (same as 74) | |
| | 66. Thrust washer | |
| | 67. Side gear | |
| | 68. Pinion (2 used) | |

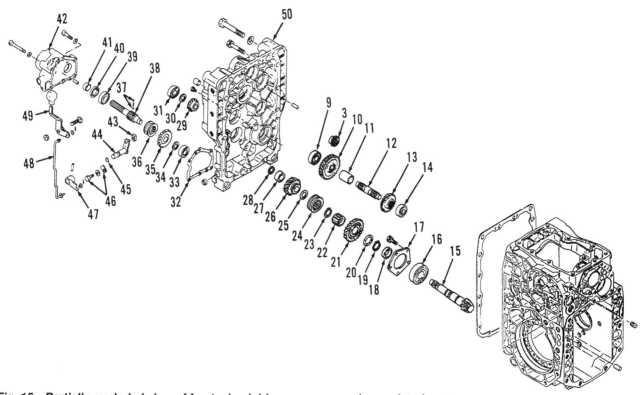

**Fig. 15—Partially exploded view of front-wheel drive gear case and associated parts.**

| | | | |
|---|---|---|---|
| 3. Drive gear | 19. Snap ring | 30. Washer | 40. Seal |
| 9. Ball bearing | 20. Washer | 31. Ball bearing | 41. Spacer |
| 10. Hi range gear | 21. Lo range gear | 32. Gasket | 42. Front-wheel drive |
| 11. Spacer | 22. Coupling hub | 33. Ball bearing | gear case |
| 12. Intermediate shaft | 23. Snap ring | 34. Washer | 43. Shift shoe |
| 13. Lo range gear | 24. Sliding coupling | 35. Driven gear | 44. Internal shift lever |
| 14. Ball bearing | 25. Washer | 36. Sliding coupling | 45. "O" ring |
| 15. Bevel pinion | 26. Hi range gear | 37. Detent balls & spring | 46. Retainer plate & screw |
| 16. Ball bearing | 27. Spacer | 38. Front-wheel drive | 47. Lever |
| 17. Retainer | 28. Snap ring | output shaft | 48. Rod |
| 18. Spacer | 29. Front-wheel drive gear | 39. Ball bearing | 49. Engaging lever |

Install shims (75) that were removed when installing differential. Make sure that bearing (74) and ring gear (71) are both firmly seated on differential case (73), then install differential in housing (7). Install bevel pinion (62), bearing (61), retainer (60), gear (59) and snap ring (58). Make sure that notch of retainer (60) is correctly positioned, then tighten screws to 26 N·m (19 ft.-lbs.) torque. Backlash between ring gear (71) and pinion (62) should be 0.17-0.23 mm (0.007-0.009 in.) when measured at outer edge of ring gear teeth. If backlash is incorrect, remove pinion and retainer assembly (58 through 62), lift differential from housing (7), then change thickness of shims (75).

When proper thickness of shims (75) has been selected, reassembly can be completed. Install new seal ring (64) and attach left axle housing (5) to differential and right axle housing (7). Apply medium strength thread sealer to retaining screws and tighten in a cross pattern to 52 N·m (38 ft.-lbs.) torque.

Refer to paragraph 15 and install shafts (55 and 62) and differential input cover (49). Install both left and right drive shafts (48 and 63), then install both final drives and steering spindles as outlined in paragraphs 13 and 14. Refer to paragraph 12 and install the axle assembly.

## FRONT WHEEL DRIVE GEAR CASE

### Four-Wheel-Drive Models

**17. R&R AND OVERHAUL.** Power for front wheel drive is from a gear (29—Fig. 15) located on front of rear drive bevel pinion shaft. To remove front drive gear case (42), first refer to paragraph 109 and drain oil from transmission, rear axle center housing and hydraulic system. Remove front axle drive shaft, remove the five screws attaching gear case housing

(42) to rear axle center housing, then remove front drive gear case and output shaft (38).

Front seal (40) must be removed and installed from inside case, with spring loaded lip toward inside.

When assembling, reverse disassembly procedure. Tighten retaining screws to 26 N·m (19 ft.-lbs.) torque. Fill transmission, rear axle center housing and hydraulic system as outlined in paragraph 109.

# ENGINE
# (MODELS 655, 755 AND 756)

**18.** Yanmar, three-cylinder, diesel engines are used on all models. Model 655 tractors are equipped with 3TN66UJ engine. Models 755 and 756 tractors are equipped with 3TNA72UJ engine.

## MAINTENANCE

## 655, 755 And 756 Models

**19. LUBRICATION.** Recommended engine lubricant is good quality engine oil with API classification CD/SF or CD/SE. Select oil viscosity depending on the ambient temperature. Recommended oil viscosities for ambient temperature ranges are as follows.

Arctic Oil . . . . . . . . . . . . . . . . . . . . . . . −55° to 0° C
(−67° to 32° F)
SAE 5W30 . . . . . . . . . . . . . . . . . . . . −30° to +10° C
(−22° to +50° F)
SAE 10W . . . . . . . . . . . . . . . . . . . . . −20° to +10° C
(−4° to +50° F)
SAE 10W30 . . . . . . . . . . . . . . . . . . . −20° to +20° C
(−4° to +68° F)
SAE 15W30 . . . . . . . . . . . . . . . . . . . −15° to +30° C
(+5° to +86° F)
SAE 15W40 . . . . . . . . . . . . . . . . . . . −15° to +40° C
(+5° to +104° F)
SAE 20W40 . . . . . . . . . . . . . . . . . . . −10° to +40° C
(+14° to +104° F)
SAE 30 . . . . . . . . . . . . . . . . . . . . . . . . 0° to +40° C
(+32° to +104° F)
SAE 40 . . . . . . . . . . . . . . . . . . . . . . +10° to +50° C
(+50° to +122° F)

Drain engine oil and remove oil filter, then install new filter and refill with new oil after the first 50 hours of operation for new or rebuilt engine. Oil and filter should be changed every 200 hours of normal operation.

Capacity of engine crankcase, including filter, is 2.4 L (2.5 qt.) for 655 model, 3.2 L (3.4 qt.) for 755 and 756 models. Oil should be maintained between marks on dipstick of all models.

**20. AIR FILTER.** The air filter should be cleaned after every 200 hours of operation or more frequently if operating in extremely dusty conditions. Remove dust from filter element by tapping lightly with your hand. Low pressure air (less than 210 kPa or 30 psi) can be directed through filter from the inside, toward outside, to blow dust from filter. Be careful not to damage filter element while attempting to clean. A new element should be installed if condition of old element is questionable. If special air cleaner solution and water are used to wash air cleaner, make sure element is dry before installing.

NOTE: Do not wash filter with any petroleum solvent.

**21. WATER PUMP/FAN/ALTERNATOR BELT.** Fan belt tension should be maintained so belt will deflect approximately 13 mm (½ in.) when moderate thumb pressure is applied to belt midway between pump and alternator pulleys. Reposition alternator to adjust belt tension. Belt length is 876 mm (34.5 in.).

**22. FUEL FILTER.** A renewable paper filter element type fuel filter is located between fuel feed pump and fuel injection pump. Check the clear filter sediment bowl daily for evidence of water or sediment. Bowl should be drained and new filter installed at least every 200 hours of operation or if filter appears dirty. Air must be bled from filter and fuel system as outlined in paragraph 82 each time filter is removed.

## R & R ENGINE ASSEMBLY

## 655, 755 And 756 Models

**23.** To remove the engine from tractor, remove grille, engine side panels, hood, battery, fuel tank, oil cooler, and radiator. Remove the muffler, pedestal side panels and the starting aid module. Detach fuel line, throttle cable, throttle cable housing, and tachometer cable. Disconnect wires to fuel shut-off, fuel transfer pump, glow plug, alternator, oil pressure sending unit, temperature sender, and starter motor at electrical connectors or terminals. Detach clamp holding wiring harness to top of engine, then move wiring and cables out of the way. Unbolt front of drive shaft from engine and attach hoist to the two lifting eyes of engine. Remove the four engine mounting bolts and lift engine from tractor frame. Be careful to remove, identify and save any shims located at engine mounts for reinstallation at same location.

Reassemble by reversing removal procedure. Two shims should be located under each front engine mount of 655 tractors. Clearance between fan blades and radiator must be at least 12 mm (½ in.) for all models. Center to center distance between side panel mounting tabs should be 508 mm (20 in.) for 655 models; 562 mm (22.1 in.) for 755 and 756 models. Tighten drive shaft attaching screws to 49 N•m (36 ft.-lbs.) torque. Refer to paragraph 82 for bleeding the fuel system.

## ENGINE TO TRANSMISSION DRIVE SHAFT

### 655, 755 And 756 Models

**24. REMOVE AND REINSTALL.** To remove the drive shaft connecting engine to hydrostatic transmission, first park tractor and block wheels to prevent movement. Turn depth control lever clockwise until it stops, then remove the four screws retaining panel located between instrument pedestal and seat. Lift right rear corner of panel first, followed by left side until clear of depth control bolt, then move panel

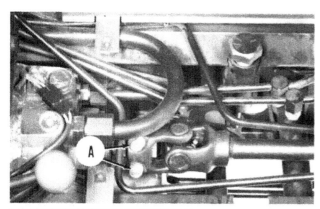

*Fig. 16—Views of drive shaft (D) connecting engine to transmission. Two clamp screws (A) are used at rear, three screws (B) attach front of drive shaft to isolation coupling and three screws (C) attach isolation coupling to engine.*

to the right and remove from tractor. Loosen the two clamp screws (A—Fig. 16) of drive shaft rear coupling, then remove the six screws (B and C) at front of drive shaft. Three screws (B) attach drive shaft to coupling and three screws (C) attach coupling to engine. Coupling isolator can be removed from between drive shaft and engine after the six screws are removed and drive shaft is moved to rear.

Slide drive shaft onto transmission input shaft, then position isolator coupling between front of drive shaft and engine. Raised part of isolator coupling, around the three unthreaded holes, should be toward rear and aligned with the three engine attaching holes. Attach isolator coupling to engine with the three longer screws, then attach drive shaft to isolator coupling with the three shorter screws. Tighten all six screws to 49 N•m (36 ft.-lbs.) torque. Tighten the two bolts (A) which clamp drive shaft rear universal joint to transmission to 60 N•m (45 ft.-lbs.) torque. Install panel between seat and instrument pedestal, being careful to align depth control lever with adjusting bolt head.

## CYLINDER HEAD

### 655, 755 And 756 Models

**25. REMOVE AND REINSTALL.** To remove cylinder head, drain coolant and engine oil. Remove water pump, rocker arm cover, rocker arm assembly, valve caps and push rods. Remove fuel injection lines and fuel return lines. The inlet manifold, exhaust manifold, glow plugs and injection nozzles do not need to be removed to remove cylinder head, but if required for other service, these parts should be removed before removing head. Remove retaining screws then lift cylinder head from engine.

Refer to appropriate paragraphs for servicing valves, guides and springs. Clean cylinder block and cylinder head gasket surfaces and inspect cylinder head for cracks, distortion or other defects. If alignment pins were removed with cylinder head, remove pins and install in cylinder block. Check head gasket surface with a straightedge and feeler gauge. Recondition or renew cylinder head if distortion exceeds 0.15 mm (0.006 in.). Cylinder head thickness should not be reduced more than 0.20 mm (0.008 in.).

Place cylinder head gasket over alignment pins and make sure that oil passage is over passage in block. Lubricate bolts with engine oil and install bolts, tightening in three steps in the sequence shown in Fig. 17. Refer to the following for correct three step torques.

**655**

| | |
|---|---|
| Engine model | 3TN66UJ |
| First torque | 10.5 N•m (7.8 ft.-lbs.) |
| Second torque | 21.1 N•m (15.6 ft.-lbs.) |
| Third (final) torque | 34 N•m (25 ft.-lbs.) |

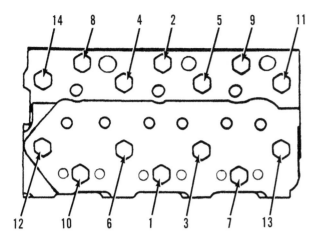

*Fig. 17—Cylinder head retaining screws should be tightened in at least three stages in the sequence shown. Refer to text for recommended torque values for each stage.*

### 755 and 756

Engine model . . . . . . . . . . . . . . . . . . . . 3TNA72UJ
First torque . . . . . . . . . . . . . 18.9 N•m (13.5 ft.-lbs.)
Second torque . . . . . . . . . . . 37.2 N•m (27.4 ft.-lbs.)
Third (final) torque . . . . . . . . . . 61 N•m (45 ft.-lbs.)

Tighten stud nuts retaining rocker arm assembly to 26 N•m (19 ft.-lbs.) torque. Tighten rocker arm cover retaining nuts to 18 N•m (13 ft.-lbs.) torque. Tighten injector nozzle to cylinder head to 50 N•m (37 ft.-lbs.) torque and nozzle leak-off line nuts to 40 N•m (30 ft.-lbs.) torque. Tighten inlet manifold retaining screws to 11 N•m (96 in.-lbs.) torque. Tighten exhaust manifold retaining nuts and screws to 11 N•m (96 in.-lbs.) torque for 655 models; 26 N•m (19 ft.-lbs.) torque for 755, 756 models. Tighten water pump retaining screws to 26 N•m (19 ft.-lbs.) torque.

## VALVE CLEARANCE

### 655, 755 And 756 Models

**26.** Valve clearance should be adjusted with engine cold. Specified clearance is 0.2 mm (0.008 in.) for both inlet and exhaust valves. Adjustment is make by loosening locknut and turning adjusting screw at push rod end of rocker arm.

To adjust valve clearance, turn crankshaft in normal direction of rotation so that rear (number 1) piston is at top dead center of compression stroke. Adjust clearance of inlet and exhaust valves for rear cylinder. Turn crankshaft 240° until next piston in firing order (front) is at top dead center and adjust clearance for both valves of that cylinder. Turn crankshaft another 240° until last piston in firing order (center) is at top dead center, then adjust valve clearance for both valves of the last cylinder.

## VALVES, GUIDES AND SEATS

### 655, 755 And 756 Models

**27.** Inlet and exhaust valves seat directly in cylinder head; however, seat inserts (3 and 5—Fig. 18) are available for service. Inlet valve face and seat angle is 30° and exhaust valve face and seat angle is 45°. Refer to the following recommended valve, guide and seat specifications.

**655**
Engine model . . . . . . . . . . . . . . . . . . . . . 3TN66UJ

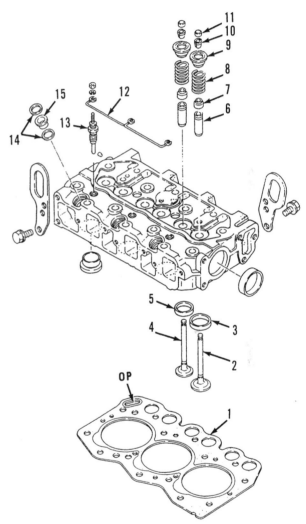

*Fig. 18—Exploded view of cylinder head used on 755 and 756 models. Parts used on 655 are similar. Oil passage is located at (OP).*

1. Gasket
2. Inlet valve
3. Valve seat
4. Exhaust valve
5. Valve seat
6. Valve guide
7. Valve stem seal
8. Valve spring
9. Spring seat
10. Split retainer
11. Valve cap
12. Glow plug connector
13. Glow plug
14. Gaskets
15. Seal

Inlet valve seat width –
  Recommended . . . . . . . . . . . 1.15 mm (0.045 in.)
  Maximum limit . . . . . . . . . . 1.65 mm (0.065 in.)
Exhaust valve seat width –
  Recommended . . . . . . . . . . . 1.41 mm (0.056 in.)
  Maximum limit . . . . . . . . . . 1.91 mm (0.075 in.)
Inlet valve seat recession –
  Recommended . . . . . . . . . . . 0.40 mm (0.016 in.)
  Maximum limit . . . . . . . . . . 0.50 mm (0.020 in.)
Exhaust valve seat recession –
  Recommended . . . . . . . . . . . 0.85 mm (0.034 in.)
Inlet and exhaust valve
  stem OD . . . . . . . . . . . . . . . 5.40 mm (0.213 in.)
Inlet and exhaust valve
  guide ID . . . . . . . . . . . . . . . 5.58 mm (0.220 in.)
Valve guide height. . . . . . . . . 7.00 mm (0.276 in.)

**755 and 756**
Engine model. . . . . . . . . . . . . . . . . 3TNA72UJ
Inlet valve seat width –
  Recommended . . . . . . . . . . . 1.44 mm (0.057 in.)
  Maximum limit . . . . . . . . . . 1.98 mm (0.078 in.)
Exhaust valve seat width –
  Recommended . . . . . . . . . . . 1.77 mm (0.070 in.)
  Maximum limit . . . . . . . . . . 2.27 mm (0.089 in.)
Inlet valve seat recession –
  Recommended . . . . . . . . . . . 0.50 mm (0.020 in.)
Exhaust valve seat recession –
  Recommended . . . . . . . . . . . 0.85 mm (0.034 in.)
Inlet and exhaust valve
  stem OD . . . . . . . . . . . . . . . 6.90 mm (0.272 in.)
Inlet and exhaust valve
  guide ID . . . . . . . . . . . . . . . 7.08 mm (0.279 in.)
Valve guide height. . . . . . . . . 9.00 mm (0.354 in.)

Valve stem to guide clearance should not exceed
0.15 mm (0.006 in.); however, guides can be knurled
to reduce clearance if less than 0.20 mm (0.008 in.).
Install new guide if clearance exceeds 0.20 mm (0.008
in.) with new valve. Press new guide (6) into cylinder
head until height of guide above surface of cylinder
head is correct as listed in preceding specifications.
Valve seat width can be narrowed using 60° and 30°
stones for exhaust valve seats; 45° and 15 ° stones for
inlet valve seats.

## VALVE SPRINGS

### 655, 755 And 756 Models

**28.** Valve springs (8—Fig. 18) are interchangeable
for inlet and exhaust valves. Renew springs which are
distorted, discolored by heat or fail to meet the following test specifications.

**655**
Engine model. . . . . . . . . . . . . . . . . 3TN66UJ
Valve spring free length . . . . . . 27.5 mm (1.083 in.)
Pressure at compressed height –
        125 N @ 17 mm (28 lbs. @ 0.591 in.)

**755 and 756**
Engine model. . . . . . . . . . . . . . . . . 3TNA72UJ
Valve spring free length . . . . . . 36.9 mm (1.453 in.)
Pressure at compressed height –
        299 N @ 22.5 mm (67 lbs. @ 0.886 in.)

## ROCKER ARMS AND PUSH RODS

### 655, 755 And 756 Models

**29.** The rocker arm assembly (Fig. 19) can be unbolted and removed after removing rocker arm cover.
Push rods can be withdrawn after removing the
rocker arms. Rocker arms are not interchangeable
and must be assembled to position ends correctly over
valve and push rod. All valve train components
should be identified as removed so they can be reinstalled in original locations if reused. Install new
component if old part does not meet standard listed
in the following specifications.

**655**
Engine model. . . . . . . . . . . . . . . . . 3TN66UJ
Rocker shaft OD, Min. . . . . . . . . 9.96 mm (0.392 in.)
Shaft support ID, Max. . . . . . . 10.09 mm (0.397 in.)
Rocker arm ID, Max. . . . . . . . . 10.09 mm (0.397 in.)
Rocker arm to shaft
  clearance, Max. . . . . . . . . . . . 0.13 mm (0.005 in.)
Push rod runout, Max. . . . . . . . . 0.30 mm (0.012 in.)
Push rod length, Min. . . . . . . . 114.0 mm (4.49 in.)

**755 and 756**
Engine model. . . . . . . . . . . . . . . . . 3TNA72UJ
Rocker shaft OD, Min. . . . . . . . 11.96 mm (0.471 in.)
Shaft support ID, Max. . . . . . . 12.09 mm (0.476 in.)
Rocker arm ID, Max. . . . . . . . . 12.09 mm (0.476 in.)
Rocker arm to shaft
  clearance, Max. . . . . . . . . . . . 0.13 mm (0.005 in.)
Push rod runout, Max. . . . . . . . . 0.30 mm (0.012 in.)
Push rod length, Min. . . . . . . . 141.0 mm (5.55 in.)

Align set screw (22) in center support (23) with hole
in shaft (24), then tighten set screw. Assemble rocker

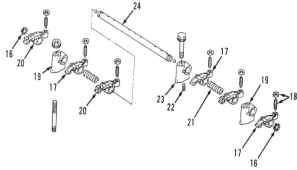

*Fig. 19—Exploded view of rocker arm and shaft assembly
typical of 655, 755 and 756 models.*

16. Snap ring
17. Inlet rocker arm
18. Adjuster screw & locknut
19. Support
20. Exhaust rocker arm
21. Spacer spring
22. Set screw
23. Center support
24. Shaft

arms, springs and remaining supports, then install snap rings (16) at ends of shaft. Tighten stud nuts and screw retaining rocker arm assembly to 26 N·m (226 in.-lbs.) torque and adjust valve clearance as outlined in paragraph 26. Tighten rocker arm cover retaining nuts to 18 N·m (160 in.-lbs.) torque.

## CAM FOLLOWERS

### 655, 755 and 756 Models

**30.** Cam followers (32—Fig. 20) can be removed from 3TN66UJ and 3TN72UJ engine models after removing cylinder head as outlined in paragraph 25. Cam followers must be identified as they are removed so that they can be reinstalled in same bores.

Minimum outside diameter of cam follower is 17.93 mm (0.706 in.) for 655 models; minimum outside diameter of cam follower for 755 and 756 models is 20.93 mm (0.824 in.). Maximum inside diameter of cam follower bore in cylinder block is 18.05 mm (0.711 in.) for 655 models; 21.05 mm (0.829 in.) for 755 and 756 models. Cam follower to bore clearance should not exceed 0.10 mm (0.004 in.) for 655 models; 0.15 mm (0.006 in.) for 755 and 756 models.

## TIMING GEARS

### 655, 755 And 756 Models

**31.** The timing gear train consists of crankshaft gear (1—Fig. 21), idler gear (2), camshaft gear (3) and

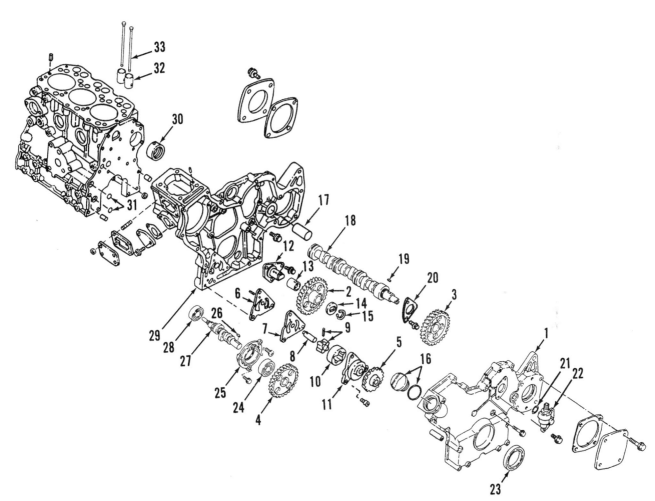

*Fig. 20—Exploded view of timing gear cover and timing gear housing typical of 655, 755 and 756 models.*

| | | |
|---|---|---|
| 1. Timing gear cover | 10. Outer rotor | 18. Valve camshaft | 26. Drive key |
| 2. Idler gear | 11. Pump housing | 19. Drive key | 27. Injection pump camshaft |
| 3. Camshaft drive gear | 12. Idler shaft | 20. Thrust plate | 28. Ball bearing |
| 4. Injection pump drive gear | 13. Idler gear bushing | 21. "O" ring | 29. Timing gear housing |
| 5. Oil pump drive gear | 14. Thrust washer | 22. Tachometer gear unit | 30. Camshaft bearing |
| 6. Gasket | 15. Snap ring | 23. Crankshaft front seal | 31. "O" rings |
| 7. Oil pump cover | 16. Filler cap & gasket | 24. Ball bearing | 32. Cam follower |
| 8. Pump shaft | 17. Plug | 25. Retainer | 33. Push rod |
| 9. Inner rotor & drive pin | | | |

*Fig. 21—View of timing gears typical of 655, 755 and 756 models. Timing marks for 3TN66UJ engine in 655 models are slightly different than similar timing marks for 3TNA72UJ engine in 755 and 756 models.*

fuel injection pump drive gear (4). Timing gear marks are visible after removing timing gear cover as outlined in paragraph 36. Valves and injection pump are properly timed to crankshaft when marks on gears are aligned as shown in Fig. 21. Because of the number of teeth on idler gear, timing marks will not align every other revolution of the crankshaft. Alignment can be checked by turning crankshaft until keyway is vertical, then checking location of mark on camshaft gear. If mark on camshaft gear is away from idler gear, turn crankshaft one complete turn until keyway is again vertical. Remove idler gear, then reinstall idler gear with marks on all gears aligned as shown in Fig. 21.

> NOTE: Before removing any gears, first refer to paragraph 29 and remove rocker arm assembly to avoid possible damage to piston or valve train. Damage could result if either the camshaft or crankshaft is turned independently from the other unless rocker arms are removed.

Backlash between timing gears (1, 2, 3 or 4) and any one of the other timing gears should be less than 0.20 mm (0.008 in.). Backlash between oil pump gear

(5) and crankshaft gear (1) should be less than 0.25 mm (0.010 in.).

Idler gear (2) of 3TN66UJ engine in 655 models has double timing mark located at (T1). Injection pump gear (4) is equipped with marks on two adjacent tooth valleys at (T2) and camshaft gear (3) has two adjacent marked teeth at (T3).

All timing gears (1, 2, 3 and 4) of 3TNA72UJ engine in 755 and 756 models have a single timing mark at (T1, T2 and T3). Marks (T1 and T3) on idler gear are located on teeth and mark (T2) on idler gear is located in tooth valley.

**32. IDLER GEAR AND SHAFT.** To inspect or remove the idler gear (2—Fig. 21) and shaft (12—Fig. 20), first remove timing gear cover as outlined in paragraph 36. Check backlash between idler gear and other timing gears before removing the gear. Backlash between idler gear and any other gear (1, 3 or 4—Fig. 21) should be less than 0.20 mm (0.008 in.).

> NOTE: Before removing any gears, first refer to paragraph 29 and remove rocker arm assembly to avoid possible damage to piston or valve train. Damage could result if either camshaft or crankshaft is turned independently from the other unless rocker arms are removed.

Remove snap ring (15—Fig. 20) from end of idler shaft, then withdraw gear from idler shaft. Idler shaft (12) can be unbolted from front face of cylinder block. Inside diameter of bushing (13) should be 20.00-20.021 mm (0.786-0.788 in.), but should not exceed 20.08 mm (0.761 in.). Outside diameter of shaft should be 19.959-19.980 mm (0.786-0.787 in.), but should not be less than 19.93 mm (0.785 in.). Renew shaft and/or bushing inside gear if clearance exceeds 0.15 mm (0.006 in.). Oil hole in bushing must be aligned with hole in idler gear. Inspect gear teeth for wear or scoring.

To install idler gear, position crankshaft with rear (number 1) piston at top dead center. Turn camshaft gear (3—Fig. 21) and fuel injection pump drive gear (4) so that timing marks point to center of the idler gear shaft. Install idler gear (2) so that all timing marks are aligned as shown in Fig. 21, then install idler gear retaining snap ring.

**33. CAMSHAFT GEAR.** The camshaft must be removed as outlined in paragraph 38 before gear (3—Fig. 20) can be pressed from shaft. Check camshaft end play and backlash between camshaft gear (3—Fig. 21) and idler gear (2) before removing camshaft. Backlash should be 0.04-0.12 mm (0.0016-0.0047 in.), but should not exceed 0.20 mm (0.008 in.). Camshaft end play should be 0.05-0.15 mm (0.002-0.006 in.), but should not exceed 0.4 mm (0.016 in.). If end play is excessive, install new thrust plate (20—Fig. 20).

Remove rocker arm assembly as outlined in paragraph 29 and idler gear as outlined in paragraph 32. Damage could result if either camshaft or crankshaft is turned independently from the other unless rocker arms are removed. Remove camshaft as outlined in paragraph 38. Gear can be pressed from camshaft to renew thrust plate (20) or gear (3).

When reinstalling camshaft in all models, tighten screws retaining thrust plate to front of engine to 11 N·m (96 in.-lbs.) torque. Refer to paragraph 29 for installation of rocker arm assembly, paragraph 36 for installation of timing gear cover and to paragraph 26 for adjusting valve clearance.

**34. CRANKSHAFT GEAR.** Crankshaft gear (1—Fig. 21) should not be removed unless a new gear is to be installed. Backlash between crankshaft gear (1) and idler gear (2) should be 0.11-0.19 mm (0.0043-0.0075 in.), but should not exceed 0.20 mm (0.008 in.). Backlash between crankshaft gear (1) and oil pump gear (5) should be 0.11-0.19 mm (0.0043-0.0075 in.), but should not exceed 0.20 mm (0.008 in.).

To remove crankshaft gear (1), the crankshaft should be removed and gear pressed from shaft.

**35. INJECTION PUMP DRIVE GEAR.** Refer to paragraph 88 for removal or other service procedures to the injection pump drive gear (4—Fig. 20).

## TIMING GEAR COVER AND CRANKSHAFT FRONT OIL SEAL

### 655, 755 And 756 Models

**36.** Crankshaft front oil seal (23—Fig. 20) can be removed and new seal installed after removing crankshaft pulley. Seal should be driven into cover until flush with front of cover. Lubricate seal before installing crankshaft pulley.

To remove timing gear cover (1), first refer to paragraph 23 and remove engine. Remove fan and alternator. Remove retaining screw from center of crankshaft pulley. Note that screw in center of crankshaft pulley should have "Loctite" or equivalent on threads and may be difficult to loosen. Use an appropriate puller and remove crankshaft pulley from crankshaft. Unbolt and remove timing gear cover from front of engine

Clean mating surfaces of timing gear cover and timing gear housing, then apply an even bead of sealer to mating surface. A gasket is not used between cover and housing. Install cover and tighten retaining screws to 9 N·m (78 in.-lbs.) torque. The crankshaft pulley of some models is equipped with a pin which must engage a hole in crankshaft gear. Coat threads of center screw with medium strength "Loctite" or equivalent and tighten to 115 N·m (85 ft.-lbs.) torque. Remainder of assembly is reverse of disassembly.

## TIMING GEAR HOUSING

### 655, 755 And 756 Models

**37.** To remove timing gear housing (29—Fig. 20), first remove engine as outlined in paragraph 23, rocker arm cover and rocker arm assembly as outlined in paragraph 29, the timing gear cover (1) as outlined in paragraph 36, camshaft and drive gear (3 and 18) as outlined in paragraph 38, and injection pump camshaft and drive gear (4 and 27) as outlined in paragraph 88. Unbolt and remove oil pump (5 through 11) from timing gear housing and oil pan (sump) from bottom of engine, then unbolt timing gear housing (29) from front of cylinder block.

When reinstalling timing gear housing, reverse removal procedure. Clean and dry sealing surfaces of timing gear housing and cylinder block, then apply bead of RTV sealer on block mating surface of timing gear housing. Install timing gear housing and tighten retaining screws to 9 N·m (78 in.-lbs.) torque.

## CAMSHAFT

### 655, 755 And 756 Models

**38.** To remove the engine camshaft, first remove rocker arm cover and rocker arm assembly as outlined in paragraph 29 and timing gear cover as outlined in paragraph 36. Check camshaft end play and backlash between camshaft gear (3—Fig. 21) and idler gear (2) before removing camshaft. Backlash should be 0.04-0.12 mm (0.0016-0.0047 in.), but should not exceed 0.20 mm (0.008 in.). Camshaft end play should be 0.05-0.15 mm (0.002-0.006 in.), but should not exceed 0.4 mm (0.16 in.). If end play is excessive, install new thrust plate (20—Fig. 20). Remove and invert engine to hold cam followers (32) away from camshaft (18).

> NOTE: It may not be necessary to turn engine up-side-down if magnets are available to hold all of the cam followers away from camshaft.

Remove camshaft thrust plate retaining screws through the web holes of camshaft gear (3), then pull camshaft (18) out of engine block bores. Gear can be pressed from camshaft to renew thrust plate (20) or gear. Flywheel must be removed to remove plug from rear of camshaft bore in cylinder block. Refer to the following specifications.

**655**
Engine model . . . . . . . . . . . . . . . . . . . . . 3TN66UJ
Camshaft journal OD —
   Front . . . . . . . . . . . . . . . . . . . . . 35.94-35.96 mm
                           (1.415-1.416 in.)
   Wear limit . . . . . . . . . . . . . . . . . . . . . 35.85 mm
                           (1.411 in.)

Intermediate . . . . . . . . . . . . . . . 35.91-35.94 mm
(1.414-1.415 in.)
    Wear limit . . . . . . . . . . . . . . . . . . . 35.81 mm
(1.410 in.)
Rear . . . . . . . . . . . . . . . . . . . . 35.94-35.96 mm
(1.415-1.416 in.)
    Wear limit . . . . . . . . . . . . . . . . . . . 35.85 mm
(1.411 in.)
Camshaft front bearing ID . . . . . 36.00-36.065 mm
(1.417-1.420 in.)
    Wear limit . . . . . . . . . . . . . . . . . . . 36.1 mm
(1.421 in.)
Camshaft end play . . . . . . . . . . . . . . 0.05-0.15 mm
(0.002-0.006 in.)
    Wear limit . . . . . . . . . . . . . . . . . . . 0.4 mm
(0.16 in.)
Cam lobe height . . . . . . . . . . . . . . 29.97-30.03 mm
(1.180-1.182 in.)
    Wear limit . . . . . . . . . . . . . . . . . . . 29.7 mm
(1.169 in.)

**755 and 756**
Engine model . . . . . . . . . . . . . . . . . . . 3TNA72UJ
Camshaft journal OD —
Front . . . . . . . . . . . . . . . . . . . . 39.94-39.96 mm
(1.572-1.573 in.)
    Wear limit . . . . . . . . . . . . . . . . . . . 39.85 mm
(1.569 in.)
Intermediate . . . . . . . . . . . . . 39.910-39.935 mm
(1.571-1.572 in.)
    Wear limit . . . . . . . . . . . . . . . . . . . 39.85 mm
(1.569 in.)
Rear . . . . . . . . . . . . . . . . . . . . 39.94-39.96 mm
(1.572-1.573 in.)
    Wear limit . . . . . . . . . . . . . . . . . . . 39.85 mm
(1.569 in.)
Camshaft front bearing ID . . . . . . 40.0-40.065 mm
(1.575-1.577 in.)
    Wear limit . . . . . . . . . . . . . . . . . . . 40.10 mm
(1.579 in.)
Camshaft end play . . . . . . . . . . . . . . 0.05-0.15 mm
(0.002-0.006 in.)
    Wear limit . . . . . . . . . . . . . . . . . . . 0.4 mm
(0.16 in.)
Cam lobe height . . . . . . . . . . . . . . 33.95-34.05 mm
(1.337-1.341 in.)
    Wear limit . . . . . . . . . . . . . . . . . . . 33.75 mm
(1.329 in.)

Be sure to align oil hole in bushing with passage in cylinder block if new bushing (30—Fig. 20) is installed.

When reinstalling camshaft in all models, tighten screws retaining thrust plate to front of engine to 11 N•m (96 in.-lbs.) torque. Refer to paragraph 29 for installation of rocker arm assembly, paragraph 36 for installation of timing gear cover and to paragraph 26 for adjusting valve clearance.

## ROD AND PISTON UNITS

### 655, 755 And 756 Models

**39.** Pistons and connecting rod units are removed from above after removing oil pan and cylinder head. Unbolt and remove oil pump suction tube from 755 and 756 models. Be sure that ridge (if present) is removed from top of cylinder bore before removing pistons. If not already marked, stamp cylinder number on piston, rod and cap before removing. Number "1" cylinder is at rear and connecting rods should be marked on injection pump (right) side. Keep rod bearing cap with matching connecting rod.

Piston must be assembled to connecting rod with recessed slot (S—Fig. 22) in top of piston opposite slot in connecting rod for bearing insert tang (T). Screws attaching connecting rod cap to rod should not be reused. Always install **new** connecting rod screws (12).

Lubricate cylinder, crankpin and ring compressor with clean engine oil. Space piston ring end gaps evenly around piston, but not aligned with piston pin, then compress piston rings with suitable compressor tool. Install piston and rod units in cylinder block with recessed slot (S) in top of piston on same side as fuel injection pump (right). Make sure that bore in connecting rod and cap is completely clean and install bearing inserts with tangs engaging slot in rod and cap. Make sure that inserts are firmly seated, lubricate crankpin and bearing inserts, then seat connecting rod with insert against crankpin. Install connecting rod cap with insert firmly seated and lubricated. Tangs on both bearing inserts should be on left (camshaft) side of engine. Tighten connecting rod cap retaining screws to 23 N•m (200 in.-lbs.) torque. On 755 and 756 models, install new seal and oil pump pickup tube, then tighten retaining screws to 11 N•m (96 in.-lbs.) torque. On all models, clean and dry sealing surface of oil pan and cylinder block, then coat sealing surface with RTV sealer and install oil pan. Tighten screws retaining oil pan to cylinder block to 11 N•m (96 in.-lbs.) torque and screws retaining oil pan to timing gear housing to 9 N•m (78 in.-lbs.) torque. Remainder of assembly is reverse of disassembly.

Clearance between flat top of piston and cylinder head can be measured using 10 mm (0.4 in.) long sections of 1.5 mm (0.06 in.) diameter soft lead wire positioned on top of the piston. Turn crankshaft one complete revolution, then remove lead wire and measure thickness of flattened wire. Clearance should be 0.59-0.74 mm (0.023-0.029 in.) for 655 models; 0.61-0.79 mm (0.024-0.031 in.) for 755 and 756 models.

# PISTON, RINGS AND CYLINDER

## 655, 755 And 756 Models

**40.** Pistons are fitted with two compression rings and one oil control ring. Piston pin is full floating and retained by a snap ring at each end of pin bore in piston. If necessary to separate piston from connecting rod, remove snap rings and push pin from piston and connecting rod. Service pistons are available in standard size and 0.25 mm (0.010 in.) oversize.

Inspect all piston skirts for scoring, cracks and excessive wear. On 655 models, measure piston skirt at right angles to pin bore 5.0 mm (0.197 in.) from bottom of skirt. Minimum skirt measurement for standard size piston is 65.85 mm (2.593 in.).

On 755 and 756 models, measure piston skirt at right angles to pin bore 8.0 mm (0.315 in.) from bottom of skirt. Minimum skirt measurement for standard size piston is 71.81 mm (2.827 in.).

Piston pin bore in piston should not exceed 20.02 mm (0.788 in.) for 655 models; 21.02 mm (0.828 in.) for 755 and 756 models. Clearance between piston pin and bore in piston should not exceed 0.05 mm (0.002 in.).

Standard cylinder bore diameter for 655 models should be 66.00-66.03 mm (2.598-2.600 in.) and should not exceed 66.20 mm (2.606 in.). Piston to cylinder bore clearance should not exceed 0.25 mm (0.010 in.). Cylinders can be rebored and fitted with 0.25 mm (0.10 in.) oversize pistons.

Standard cylinder bore diameter for 755 and 756 models should be 72.00-72.03 mm (2.835-2.836 in.) and should not exceed 72.20 mm (2.843 in.). Piston to cylinder bore clearance should not exceed 0.28 mm (0.011 in.). Cylinders can be rebored and fitted with 0.25 mm (0.10 in.) oversize pistons.

Cylinder bore of all models should be deglazed using 180 grit stone and crosshatch pattern should intersect at 30-40°. Wash abrasive from cylinder walls using warm soapy water. Continue cleaning until white cleaning cloths show no discoloration. Make sure all soap is rinsed from cylinders, dry cylinder block thoroughly, then lubricate with clean engine oil.

**NOTE: Use only mild soap to clean abrasive from cylinders. Do not use commercial solvents, gasoline or kerosene to clean cylinders.**

End gap of piston rings in cylinder should not exceed 1.50 mm (0.059 in.). Rings should be inserted squarely in cylinder using piston to push ring approximately 30 mm (1.181 in.) into cylinder.

Piston ring side clearance in grooves should not exceed 0.20 mm (0.008 in.) for all rings. Renew piston if ring side clearance is excessive.

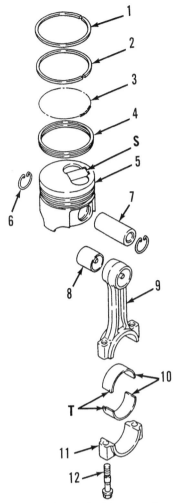

*Fig. 22—Exploded view of piston and connecting rod assembly. Assemble piston to connecting rod so that slot (S) in top of piston will be on injector side of engine and slots in connecting rod and cap for bearing insert tangs (T) will be on opposite (left) side as shown.*

| | |
|---|---|
| 1. Top compression ring | 7. Piston pin |
| 2. Second compression ring | 8. Pin bushing |
| 3. Oil ring expander | 9. Connecting rod |
| 4. Oil control ring | 10. Crankpin bearing insert |
| 5. Piston | 11. Connecting rod cap |
| 6. Retaining ring | 12. Retaining screw |

When installing piston rings, install oil ring expander in groove, followed by the oil ring. End gap of oil ring should be 180° from ends of expander. Install compression ring with inside chamfer or identification mark toward top of piston. Space end gaps of top, second and oil control rings 120° apart before installing in cylinder bore.

## PISTON PIN

## 655, 755 And 756 Models

**41.** Piston pin bore in piston should not exceed 20.02 mm (0.788 in.) for 655 models; 21.02 mm (0.828

in.) for 755 and 756 models. Clearance between piston pin and bore in piston should not exceed 0.05 mm (0.002 in.).

Piston pin diameter should be measured at each end and in center. Measure pin diameter at each of the three locations and again 90° from each of the original locations. Piston pin diameter for 655 models should not be less than 19.98 mm (0.786 in.) and pin diameter should not be less than 20.98 mm (0.826 in.) for 755 and 756 models.

Inside diameter of pin bushing in top of connecting rod should not exceed 20.10 mm (0.791 in.) for 655 models. For 755 and 756 models, pin bushing should not exceed 21.10 mm (0.831 in.). Clearance between piston pin and bore of pin bushing in connection rod should not exceed 0.11 mm (0.004 in.).

## CONNECTING ROD AND BEARINGS

### 655, 755 And 756 Models

**42.** Connecting rod can be unbolted from crankshaft crankpin and new bearing inserts installed from below after removing oil pan. Refer to paragraph 39 for removing connecting rods. Connecting rod should not be out-of-parallel or twisted more than 0.08 mm (0.003 in.).

Inside diameter of pin bushing in top of connecting rod should be 20.025-20.038 mm (0.788-0.789 in.), but should not exceed 20.10 mm (0.791 in.) for 655 models. For 755 and 756 models, pin bushing should be 21.025-21.038 mm (0.8278-0.8282 in.), but should not exceed 21.10 mm (0.831 in.). For all models, clearance between piston pin and bushing bore in connecting rod should be 0.025-0.047 mm (0.0010-0.0019 in.), but should not exceed 0.11 mm (0.004 in.).

Connecting rod bearing clearance on crankshaft crankpin should be 0.020-0.072 mm (0.0008-0.0028 in.) and wear limit should not exceed 0.15 mm (0.006 in.). Connecting rod side play on crankpin should be 0.2-0.4 mm (0.0079-0.0157 in.), but should not exceed 0.55 mm (0.0217 in.).

Crankpin standard diameter for 655 models is 35.97-35.98 mm (1.416-1.417 in.). Crankpin standard diameter for 755 and 756 models is 39.97-39.98 mm (1.5736-1.574 in.). Undersized bearing inserts, which require crankshaft journals to be resized and polished, are available for service. Finished undersize journals should provide standard clearance with undersized bearings.

Refer to paragraph 39 if connecting rod and piston units are removed from engine. Make sure that bore in connecting rod and cap is completely clean. Install bearing inserts with tangs engaging slot in rod and cap. Make sure that inserts are firmly seated, lubricate crankpin and bearing inserts, then seat connecting rod with insert against crankpin. Install connecting rod cap with insert firmly seated and lubricated. Tangs (T—Fig. 22) on both halves of bearing should be on left (camshaft) side of engine. Tighten connecting rod cap retaining screws to 23 N·m (200 in.-lbs.) torque. Install new seal and oil pump pickup tube, then tighten retaining screws to 11 N·m (96 in.-lbs.) torque. Clean and dry sealing surface of oil pan and cylinder block, then coat sealing surface with RTV sealer and install oil pan. Tighten screws retaining oil pan to cylinder block to 11 N·m (96 in.-lbs.) torque and screws retaining oil pan to timing gear housing to 9 N·m (78 in.-lbs.) torque.

## CRANKSHAFT AND MAIN BEARINGS

### 655, 755 And 756 Models

**43.** The crankshaft is supported in four main bearings. Crankshaft end play is controlled by insert type thrust bearing (8 and 10—Fig. 23) on sides of rear (number 1) main bearing journal.

To remove crankshaft (5), remove engine from tractor as outlined in paragraph 23, rocker arm cover and rocker arm assembly as outlined in paragraph 29, timing gear cover as outlined in paragraph 36, camshaft as outlined in paragraph 38, injection pump drive gear and camshaft as outlined in paragraph 88 and timing gear housing as outlined in paragraph 37. Remove flywheel (18), then remove crankshaft rear oil seal as outlined in paragraph 44. Invert engine, then refer to paragraph 39 and detach connecting rods from crankshaft. Mark all main bearing caps for correct installation, unbolt and remove crankshaft main bearing caps, then lift crankshaft from engine. The rear main bearing is number "1."

Check all main bearing journals, thrust surfaces of rear journal and connecting rod crankpin journals for scoring or excessive wear. Main bearings and rod bearings are available in standard size and 0.25 mm (0.010 in.) undersize. Refer to the following specifications.

**655**

Engine model . . . . . . . . . . . . . . . . . . . . . . 3TN66UJ
Crankpin standard OD —
  Desired . . . . . . . . . . . . . . . . . . . 35.97-35.98 mm
                                (1.416-1.417 in.)
  Wear limit. . . . . . . . . . . . . . . . . . . . . . . 35.92 mm
                                  (1.414 in.)
Crankpin to rod bearing clearance —
  Desired . . . . . . . . . . . . . . . . . . . . 0.020-0.072 mm
                              (0.0008-0.0028 in.)
  Wear limit. . . . . . . . . . . . . . . . . . . . . . . 0.15 mm
                                  (0.006 in.)
Main journal standard OD —
  Desired . . . . . . . . . . . . . . . . . . . 39.97-39.98 mm
                                  (1.5736-1.5740 in.)
  Wear limit. . . . . . . . . . . . . . . . . . . . . . . 39.92 mm
                                  (1.572 in.)

Main journal to bearing clearance —
  Desired . . . . . . . . . . . . . . . . . 0.020-0.072 mm
    (0.0008-0.0028 in.)
  Wear limit . . . . . . . . . . . . . . . . . . 0.15 mm
    (0.006 in.)

Crankshaft end play —
  Desired . . . . . . . . . . . . . . . . . 0.095-0.266 mm
    (0.004-0.011 in.)
  Wear limit . . . . . . . . . . . . . . . . . . 0.33 mm
    (0.013 in.)

Main cap torque . . . . . . . . . . . . . . . . . . 54 N·m
    (40 ft.-lbs.)

Rear oil seal case torque —
  Seal to block . . . . . . . . . . . . . . . . . . 11 N·m
    (96 in.-lbs.)
  Oil pan to seal case . . . . . . . . . . . . . . . . 9 N·m
    (78 in.-lbs.)

**755 and 756**
Engine model . . . . . . . . . . . . . . . . . 3TNA72UJ
Crankpin standard OD —
  Desired . . . . . . . . . . . . . . 39.97-39.98 mm
    (1.5736-1.574 in.)
  Wear limit . . . . . . . . . . . . . . . . . . 39.92 mm
    (1.572 in.)

Crankpin to rod bearing clearance —
  Desired . . . . . . . . . . . . . . . . . 0.020-0.072 mm
    (0.0008-0.0028 in.)
  Wear limit . . . . . . . . . . . . . . . . . . 0.15 mm
    (0.006 in.)

Main journal standard OD —
  Desired . . . . . . . . . . . . . . 43.97-43.98 mm
    (1.731-1.732 in.)
  Wear limit . . . . . . . . . . . . . . . . . . 43.92 mm
    (1.729 in.)

Main journal to bearing clearance —
  Desired . . . . . . . . . . . . . . . . . 0.020-0.072 mm
    (0.0008-0.0028 in.)
  Wear limit . . . . . . . . . . . . . . . . . . 0.15 mm
    (0.006 in.)

Crankshaft end play —
  Desired . . . . . . . . . . . . . . . . . 0.090-0.271 mm
    (0.004-0.011 in.)
  Wear limit . . . . . . . . . . . . . . . . . . 0.33 mm
    (0.013 in.)

Main cap torque . . . . . . . . . . . . . . . . . . 79 N·m
    (58 ft.-lbs.)

Rear oil seal case torque —
  Seal to block . . . . . . . . . . . . . . . . . . 11 N·m
    (96 in.-lbs.)

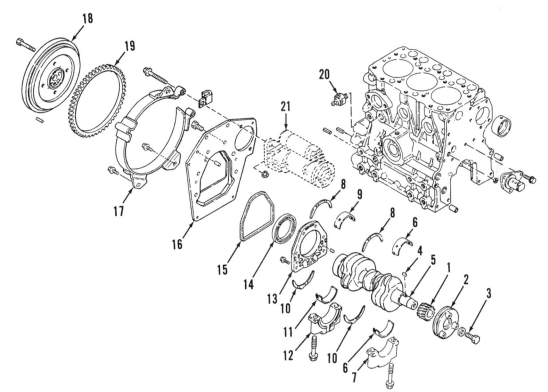

*Fig. 23—View of crankshaft, rear seal and flywheel. Model 655 is shown, but 755 and 756 are similar.*

1. Crankshaft gear
2. Crankshaft pulley
3. Retaining screw
4. Drive key
5. Crankshaft
6. Bearing insert set
7. Main bearing cap
8. Thrust bearing halves (top)
9. Rear main bearing insert (top)
10. Thrust bearing halves (bottom)
11. Rear main bearing insert (bottom)
12. Rear main bearing cap
13. Crankshaft rear seal
    retainer housing
14. Rear seal
15. Foam seal
16. Engine rear plate
17. Flywheel shield
18. Flywheel
19. Ring gear
20. Oil pressure sender
21. Starter motor

Oil pan to seal case . . . . . . . . . . . . . . . . . 9 N•m
(78 in.-lbs.)

If crankshaft gear (1—Fig. 23) was removed, heat gear to approximately 150° C (300° F) before pressing gear onto crankshaft. Be sure that all bearing surfaces are clean and free of nicks or burrs. Install bearing inserts with oil holes in the cylinder block and bearing shells without oil holes in the main bearing caps. Bearing sets for all journals are alike. Lubricate crankshaft and bearings, then set crankshaft in bearings. Lubricate thrust bearing upper halves (without projections) and insert bearing halves between block and crankshaft with oil groove side toward crankshaft shoulders. Use light grease to stick thrust washer lower halves (10) to the sides of rear main bearing cap (12) with tabs aligned with notches in cap. Install bearing insert (11) in cap, lubricate bearing and install cap in original location. Arrow on side of main bearing cap marked "FW" points to flywheel end of engine. Lubricate threads of main bearing cap retaining screws with engine oil and tighten to torque listed in specifications. Check to be sure that crankshaft turns freely before completing assembly.

## CRANKSHAFT FRONT OIL SEAL

### 655, 755 And 756 Models

**44.** The crankshaft front oil seal (23—Fig. 20) is located in timing gear cover and can be renewed after removing crankshaft pulley (2—Fig. 23). Pry seal from timing cover taking care not to damage cover. Seal lip rides on hub of crankshaft pulley. Inspect pulley hub for excessive wear at point of seal contact. Install new seal with lip toward inside using suitable driver. Lubricate seal lip and crankshaft pulley hub, install pulley on crankshaft and tighten pulley retaining screw. The crankshaft pulley of some models is equipped with a pin which must engage a hole in crankshaft gear. Coat threads of center screw (3) with medium strength "Loctite" or equivalent and tighten to 115 N•m (85 ft.-lbs.) torque.

## CRANKSHAFT REAR OIL SEAL

### 655, 755 And 756 Models

**45.** The crankshaft rear oil seal (14—Fig. 23) can be renewed after removing flywheel and seal retainer housing (13). Install new oil seal with lip toward inside. Press seal into housing until flush with outer surface of retainer housing. If crankshaft is grooved at original oil seal contact surface, new seal can be installed 3 mm (⅛ in.) farther into seal housing. Lubricate seal lip, coat sealing surface of retainer housing with RTV sealer and install retainer housing. Tighten screws attaching oil seal retainer housing to rear of cylinder block to 11 N•m (95 in.-lbs.) and

screws attaching oil pan to bottom of retainer housing to 9 N•m (78 in.-lbs.) torque.

## FLYWHEEL

### 655, 755 And 756 Models

**46.** To remove flywheel, first remove engine as outlined in paragraph 23. Unbolt and remove flywheel shield (17—Fig. 23), then unbolt and remove flywheel from rear of crankshaft. The engine rear plate (16) can be unbolted and removed from rear of engine block after removing flywheel.

When assembling, be sure to install foam seal (15) around crankshaft rear seal retainer housing (13). Tighten screws attaching engine rear plate (16) to rear of engine to 49 N•m (36 ft.-lbs.) torque and screws attaching flywheel (18) to crankshaft to 59 N•m (44 ft.-lbs.) torque. Install flywheel shield (17) and tighten screws securely.

## OIL PAN

### 655, 755 And 756 Models

**47.** Drain engine oil, then refer to paragraph 23 and remove engine. Unbolt and remove oil pan from engine. Oil pickup tube is installed in oil pan of 655 models and attached to bottom of cylinder block of 755 and 756 models. On all models, seal is installed at top of oil pump pickup tube and tube retaining screws should be tightened to 11 N•m (96 in.-lbs.) torque. Seal ring is installed around hollow dowel of 655 oil pan to provide positive seal for oil pickup tube.

Clean and dry sealing surface of oil pan and cylinder block, then coat sealing surface with RTV sealer and install oil pan. Tighten screws retaining oil pan to cylinder block to 11 N•m (96 in.-lbs.) torque and screws retaining oil pan to timing gear housing to 9 N•m (78 in.-lbs.) torque.

## OIL PUMP AND PRESSURE RELIEF VALVE

### 655, 755 And 756 Models

**48.** The gerotor type oil pump is mounted on front of engine timing gear housing and gear (5—Fig. 21) on pump shaft is driven by crankshaft gear (1).

To remove oil pump, first remove timing gear cover as outlined in paragraph 36, then measure backlash between oil pump gear (5) and crankshaft gear (1). Backlash should be 0.11-0.19 mm (0.0043-0.0075 in.), but should not exceed 0.25 mm (0.010 in.). Unbolt and remove pump.

Pump end cover (7—Fig. 20) can be removed from pump body (11) to inspect pump. End clearance of

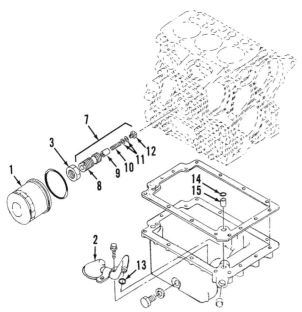

*Fig. 24—View of oil pan and relief valve used on 655 models.*

1. Oil filter
2. Oil pickup
3. Nut
7. Relief valve
8. Adapter body
9. Relief valve poppet
10. Spring
11. Adjusting shims
12. Valve retainer
13. "O" ring
14. "O" ring
15. Dowel

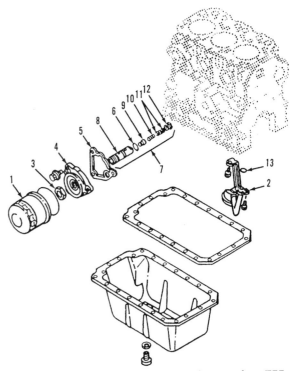

*Fig. 25—View of oil pan and relief valve used on 755 and 756 models.*

1. Oil filter
2. Oil pickup
3. Nut
4. Filter base
5. Gasket
6. "O" ring
7. Relief valve
8. Adapter body
9. Relief valve poppet
10. Spring
11. Adjusting shims
12. Valve retainer
13. "O" ring

rotors in pump body should not exceed 0.25 mm (0.10 in.). Clearance between outer rotor (10) and pump housing should not exceed 0.25 mm (0.10 in.). Clearance between tip of inner rotor (9) and center of outer rotor lobe should not exceed 0.25 mm (0.10 in.) for 655 models or 0.15 mm (0.006 in.) for 755 and 756 models. On all models, pump drive gear (5) can be pressed from shaft (8) if renewal is required, but service pump is available with drive gear. Mark on side of outer rotor (10) should be toward inside of pump housing bore.

Install pump using new gasket and tighten retaining screws to 25 N·m (18 ft.-lbs.) torque. Refer to paragraph 36 for installing timing gear housing.

Oil pressure relief valve (7—Fig. 24 or Fig. 25) is located inside threaded oil filter adapter. Relief valve pressure can be adjusted by removing valve retainer (12) and adding shims (11). Adding or removing one 1 mm (0.039 in.) thick shim will change oil pressure

about 13.8 kPa (2 psi) on 655 models or 10.9 kPa (1.6 psi) for 755 and 756 models.

Free length of spring (10) for 655 models is 21.9-24.5 mm (0.86-0.96 in.) and spring should exert 12.0 N (2.7 lbs.) when compressed to 14.7 mm (0.58 in.). Free length of spring (10) for 755 and 756 models is 43.5-48.5 mm (1.71-1.91 in.) and spring should exert 20.5 N (4.6 lbs.) when compressed to 27.5 mm (1.08 in.).

On all models, relief valve poppet (9) should move freely in bore of body (8) and should not stick. It is not necessary to remove nut (3) to service relief valve, but if nut is loosened, tighten to 30 N·m (22 ft.-lbs.) torque.

# ENGINE
## MODELS 855, 856 AND 955

**49.** Yanmar, three-cylinder, diesel engines are used on all models. Model 855 and 856 tractors are equipped with 3TN75RJ engine and Model 955 tractors are equipped with 3TN84RJ engine.

## MAINTENANCE

### 855, 856 And 955 Models

**50. LUBRICATION.** Recommended engine lubricant is good quality engine oil with API classification CD/SF or CD/SE. Select oil viscosity depending on the ambient temperature. Recommended oil viscosities for ambient temperature ranges are as follows.

| | |
|---|---|
| Arctic Oil | −55° to 0° C |
| | (−67° to 32° F) |
| SAE 5W30 | −30° to +10° C |
| | (−22° to +50° F) |
| SAE 10W | −20° to +10° C |
| | (−4° to +50° F) |
| SAE 10W30 | −20° to +20° C |
| | (−4° to +68° F) |
| SAE 15W30 | −15° to +30° C |
| | (+5° to +86° F) |
| SAE 15W40 | −15° to +40° C |
| | (+5° to +104° F) |
| SAE 20W40 | −10° to +40° C |
| | (+14° to +104° F) |
| SAE 30 | 0° to +40° C |
| | (+32° to +104° F) |
| SAE 40 | +10° to +50° C |
| | (+50° to +122° F) |

Drain engine oil and remove oil filter, then install new filter and refill with new oil after the first 50 hours of operation for new or rebuilt engine. Oil and filter should be changed every 100 hours of normal operation for 955 models; every 200 hours of normal operation for 855 and 856 models.

Capacity of engine crankcase, including filter, is 3.9 L (4.1 qt.) for 855 and 856 models, and 4.2 L (4.4 qt.) for 955 models. Oil should be maintained between marks on dipstick of all models.

**51. AIR FILTER.** The air filter should be cleaned after every 200 hours of operation or more frequently if operating in extremely dusty conditions. Remove dust from filter element by tapping lightly with your hand. Low pressure (less that 210 kPa or 30 psi) can be directed through filter from the inside, toward outside, to blow dust from filter. Be careful not to damage filter element while attempting to clean. A new element should be installed if condition of old element is questionable. If special air cleaner solution and water are used to wash air cleaner, make sure element is dry before installing.

**NOTE: Do not wash filter with any petroleum solvent.**

**52. WATER PUMP/FAN/ALTERNATOR BELT.** Fan belt tension should be maintained so belt will deflect approximately 13 mm (½ in.) when moderate thumb pressure is applied to belt midway between pump and alternator pulleys. Reposition alternator to adjust belt tension. Belt length is 965 mm (38 in.) for 855 and 856 models, and 927 mm (36.5 in.) for 955 models.

**53. FUEL FILTER.** A renewable paper filter element type fuel filter is located between fuel feed pump and fuel injection pump. Check the clear filter sediment bowl daily for evidence of water or sediment. Bowl should be drained and new filter installed at least every 200 hours of operation or if filter appears dirty. Air must be bled from filter and fuel system as outlined in paragraph 82 each time filter is removed.

## R & R ENGINE ASSEMBLY

### 855, 856 And 955 Models

**54.** To remove the engine from tractor, remove grille, engine side panels, hood, battery, fuel tank, oil cooler, and radiator. Remove the muffler, pedestal side panels and the starting aid module. Detach fuel line, throttle cable, throttle cable housing, and tachometer cable. Disconnect wires to fuel shut-off, fuel transfer pump, glow plug, alternator, oil pressure sending unit, temperature sender, and starter motor at electrical connectors or terminals. Detach clamp holding wiring harness to top of engine, then move wiring and cables out of the way. Unbolt front of drive shaft from engine and attach hoist to the two lifting eyes of engine. Remove the four engine mounting bolts and lift engine from tractor frame. Be careful to remove, identify and save any shims located at engine mounts for reinstallation at same location.

Reassemble by reversing removal procedure. Clearance between fan blades and radiator must be at least 12 mm (½ in.). Center to center distance between side panel mounting tabs should be 567 mm (22.3 in.). Tighten drive shaft attaching screws to 49 N·m (36 ft.-lbs.) torque. Refer to paragraph 82 for bleeding the fuel system.

## ENGINE TO TRANSMISSION DRIVE SHAFT

### 855, 856 And 955 Models

**55. REMOVE AND REINSTALL.** To remove the drive shaft connecting engine to hydrostatic transmission, first park the tractor and block wheels to prevent movement. Turn depth control lever clockwise until it stops, then remove the four screws retaining panel located between instrument pedestal and seat. Lift right rear corner of panel first, followed by left side until clear of depth control bolt, then move panel to the right and remove from tractor. Loosen the two clamp screws (A—Fig. 26) of drive shaft rear coupling, then remove the six screws (B and C) at front of drive shaft. Three screws (B) attach drive shaft to coupling and three screws (C) attach coupling to engine. Coupling isolator can be removed from between drive shaft and engine after the six screws are removed and drive shaft is moved to rear.

Slide drive shaft onto transmission, then position isolator coupling between front of drive shaft and engine. Raised part of isolator coupling around the three unthreaded holes should be toward rear and aligned with the three engine attaching holes. Attach isolator coupling to engine with the three longer screws, then attach drive shaft to isolator coupling with the three shorter screws. Tighten all six screws to 49 N·m (36 ft.-lbs.) torque. Tighten the two bolts which clamp drive shaft rear universal joint to transmission to 60 N·m (45 ft.-lbs.) torque. Install panel between seat and instrument pedestal, being careful to align depth control lever with adjusting bolt head.

## CYLINDER HEAD

### 855, 856 And 955 Models

**56. REMOVE AND REINSTALL.** To remove cylinder head, drain coolant and engine oil. Remove water pump, rocker arm cover, rocker arm assembly, valve caps and push rods. Remove fuel injection lines and fuel return lines. The inlet manifold, exhaust manifold, glow plugs and injection nozzles do not need to be removed to remove cylinder head, but if required for other service, these parts should be removed before removing head. Remove retaining screws then lift cylinder head from engine.

Refer to appropriate paragraphs for servicing valves, guides and springs. Clean cylinder block and cylinder head gasket surfaces and inspect cylinder head for cracks, distortion or other defects. If alignment pins were removed with cylinder head, remove pins and install in cylinder block. Check head gasket surface with a straightedge and feeler gauge. Recondition or renew cylinder head if distortion exceeds 0.15 mm (0.006 in.). Cylinder head thickness should not be reduced more than 0.20 mm (0.008 in.).

Place cylinder head gasket over alignment pins and make sure that oil passage is over passage in block. Lubricate bolts with engine oil and install bolts, tightening in three steps in the sequence shown in Fig. 27. Refer to the following for correct three step torques.

*Fig. 26—Views of drive shaft (D) connecting engine to transmission. Two clamp screws (A) are used at rear, three screws (B) attach front of drive shaft to isolation coupling and three screws (C) attach isolation coupling to engine.*

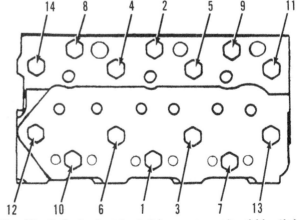

*Fig. 27—Cylinder head retaining screws should be tightened in at least three stages in the sequence shown. Refer to text for recommended torque values for each stage.*

## 855 and 856

Engine model . . . . . . . . . . . . . . . . . . . . . 3TN75RJ
First torque . . . . . . . . . . . . . . . . . . . . . . . 21.4 N•m
(15.8 ft.-lbs.)
Second torque . . . . . . . . . . . . . . . . . . . . . 42.1 N•m
(31.0 ft.-lbs.)
Third (final) torque . . . . . . . . . . . . . . . . . . 69 N•m
(51 ft.-lbs.)

## 955

Engine model . . . . . . . . . . . . . . . . . . . . . 3TN84RJ
First torque . . . . . . . . . . . . . . . . . . . . . . . 24.2 N•m
(17.8 ft.-lbs.)
Second torque . . . . . . . . . . . . . . . . . . . . . 47.6 N•m
(35.1 ft.-lbs.)
Third (final) torque . . . . . . . . . . . . . . . . . . 78 N•m
(58 ft.-lbs.)

Tighten stud nuts retaining rocker arm assembly to 26 N•m (19 ft.-lbs.) torque. Tighten rocker arm cover retaining nuts to 18 N•m (160 in.-lbs.) torque. Tighten injector nozzle to cylinder head to 50 N•m (37 ft.-lbs.) torque and nozzle leak-off line nuts to 40 N•m (30 ft.-lbs.) torque. Tighten screws retaining inlet manifold, exhaust manifold and water pump to 26 N•m (19 ft.-lbs.) torque.

## VALVE CLEARANCE

### 855, 856 And 955 Models

**57.** Valve clearance should be adjusted with engine cold. Specified clearance is 0.2 mm (0.008 in.) for both inlet and exhaust valves. Adjustment is made by loosening locknut and turning adjusting screw at push rod end of rocker arm.

To adjust valve clearance, turn crankshaft in normal direction of rotation so that rear (number 1) piston is at top dead center of compression stroke. Adjust clearance of inlet and exhaust valves for rear cylinder. Turn crankshaft 240° until next piston in firing order (front) is at top dead center and adjust clearance for both valves of that cylinder. Turn crankshaft another 240° until last piston in firing order (center) is at top dead center, then adjust valve clearance for both valves of the last cylinder.

## VALVES, GUIDES AND SEATS

### 855, 856 And 955 Models

**58.** Both inlet and exhaust valves originally seat directly in cylinder head of all except 955 models, which seat on renewable valve seat inserts. On all models, inlet valve face and seat angle is 30° and exhaust valve face and seat angle is 45°. Refer to the following recommended valve, guide and seat specifications.

## 855 and 856

Engine model . . . . . . . . . . . . . . . . . . . . . 3TN75RJ
Inlet valve seat width —
Recommended . . . . . . . . . . . . . . . . . . . . 1.44 mm
(0.057 in.)
Maximum limit . . . . . . . . . . . . . . . . . . . 1.98 mm
(0.078 in.)
Exhaust valve seat width —
Recommended . . . . . . . . . . . . . . . . . . . . 1.77 mm
(0.070 in.)
Maximum limit . . . . . . . . . . . . . . . . . . . 2.27 mm
(0.089 in.)
Inlet and exhaust valve seat recession —
Recommended . . . . . . . . . . . . . . 0.30-0.59 mm
(0.012-0.020 in.)
Inlet and exhaust valve stem OD . . . . . . . . 6.90 mm
(0.272 in.)
Inlet and exhaust valve guide ID . . . . . . . . 7.08 mm
(0.279 in.)
Valve guide height. . . . . . . . . . . . . . . . . . . 12.00 mm
(0.472 in.)

## 955

Engine model . . . . . . . . . . . . . . . . . . . . . 3TN84RJ
Inlet valve seat width —
Recommended . . . . . . . . . . . . . . . . . . . . 1.16 mm
(0.046 in.)
Maximum limit . . . . . . . . . . . . . . . . . . . 1.74 mm
(0.069 in.)
Exhaust valve seat width —
Recommended . . . . . . . . . . . . . . . . . . . . 1.35 mm
(0.053 in.)
Maximum limit . . . . . . . . . . . . . . . . . . . 1.94 mm
(0.076 in.)
Inlet and exhaust valve seat recession —
Recommended . . . . . . . . . . . . . . . . . . . . 1.80 mm
(0.071 in.)
Inlet and exhaust valve stem OD . . . . . . . . 7.90 mm
(0.311 in.)
Inlet and exhaust valve guide ID . . . . . . . . 8.10 mm
(0.319 in.)
Valve guide height. . . . . . . . . . . . . . . . . . . 15.00 mm
(0.591 in.)

Valve stem to guide clearance should not exceed 0.15 mm (0.006 in.); however, guides can be knurled to reduce clearance if less than 0.20 mm (0.008 in.). Install new guide if clearance exceeds 0.20 mm (0.008 in.) with new valve. Press new guide (6—Fig. 28) into cylinder head until height of guide above surface of cylinder head is correct as listed in preceding specifications. Valve seat width can be narrowed using 60° and 30° stones for exhaust valve seats; 45° and 15° stones for inlet valve seats.

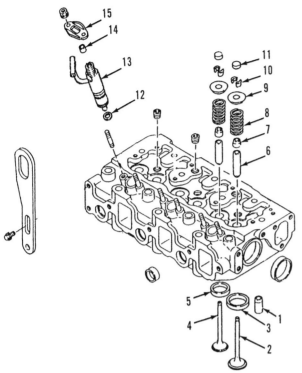

*Fig. 28—Exploded view of cylinder head used on 955 models. Parts used on 855 and 856 models are similar.*

1. Alignment pin
2. Inlet valve
3. Valve seat
4. Exhaust valve
5. Valve seat
6. Valve guide
7. Valve stem seal
8. Valve spring
9. Spring seat
10. Split retainer
11. Valve cap
12. Gasket
13. Injection nozzle
14. Seal
15. Retainer

## VALVE SPRINGS

### 855, 856 And 955 Models

**59.** Valve springs are interchangeable for inlet and exhaust valves. Renew springs which are distorted, discolored by heat or fail to meet the following test specifications.

**855 and 856**
Engine model . . . . . . . . . . . . . . . . . . . . . . 3TN75RJ
Valve spring free length . . . . . . . . . . . . . . 41.5 mm
(1.634 in.)

Pressure at
compressed height . . . . . . . . . . 313 N @ 25.2 mm
(70 lbs. @ 0.992 in.)

**955**
Engine model . . . . . . . . . . . . . . . . . . . . . 3TN84RJ
Valve spring free length . . . . . . . . . . . . . . 41.5 mm
(1.634 in.)

Pressure at
compressed height . . . . . . . . . . . 319 N @ 24 mm
(72 lbs. @ 0.945 in.)

## ROCKER ARMS AND PUSH RODS

### 855, 856 And 955 Models

**60.** The rocker arm assembly (Fig. 29) can be unbolted and removed after removing the rocker arm cover. Push rods can be withdrawn after removing rocker arms. Inlet and exhaust rocker arms are alike, but all valve train components should be identified as

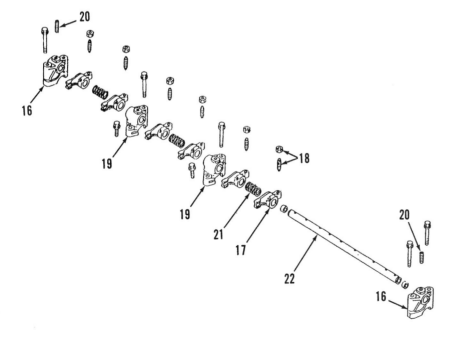

*Fig. 29—Exploded view of rocker arm and shaft assembly typical of 855, 856 and 955 models. Set screws (20) may be installed in center supports of some models.*

16. End support
17. Rocker arm
18. Adjuster screw & locknut
19. Center support
20. Set screw
21. Spacer spring
22. Shaft

they are removed so they can be reinstalled in original locations if reused. Install new component if old part does not meet standard listed in the following specifications.

**855 and 856**

| | |
|---|---|
| Engine model | 3TN75RJ |
| Rocker shaft OD — | |
| Desired | 15.966-15.984 mm |
| | (0.6286-0.6293 in.) |
| Wear limit | 15.96 mm |
| | (0.628 in.) |
| Shaft support ID — | |
| Desired | 16.0-16.02 mm |
| | (0.630-0.631 in.) |
| Wear limit | 16.09 mm |
| | (0.633 in.) |
| Rocker arm ID — | |
| Desired | 16.0-16.02 mm |
| | (0.630-0.631 in.) |
| Wear limit | 16.09 mm |
| | (0.633 in.) |
| Rocker arm to shaft clearance — | |
| Desired | 0.016-0.054 mm |
| | (0.0006-0.0021 in.) |
| Wear limit | 0.13 mm |
| | (0.005 in.) |
| Push rod runout, Max. | 0.30 mm |
| | (0.012 in.) |
| Push rod length, Min. | 146.6 mm |
| | (5.77 in.) |

**955**

| | |
|---|---|
| Engine model | 3TN84RJ |
| Rocker shaft OD — | |
| Desired | 15.966-15.984 mm |
| | (0.6286-0.6293 in.) |
| Wear limit | 15.96 mm |
| | (0.628 in.) |
| Shaft support ID — | |
| Desired | 16.0-16.02 mm |
| | (0.630-0.631 in.) |
| Wear limit | 16.09 mm |
| | (0.633 in.) |
| Rocker arm ID — | |
| Desired | 16.0-16.02 mm |
| | (0.630-0.631 in.) |
| Wear limit | 16.09 mm |
| | (0.633 in.) |
| Rocker arm to shaft clearance — | |
| Desired | 0.016-0.054 mm |
| | (0.0006-0.0021 in.) |
| Wear limit | 0.13 mm |
| | (0.005 in.) |
| Push rod runout, Max. | 0.30 mm |
| | (0.012 in.) |
| Push rod length, Min. | 1.78.2 mm |
| | (7.02 in.) |

Assemble rocker arms, springs and the two center supports. Align set screw in the two supports at ends with holes in shaft, then tighten set screws. Tighten stud nuts retaining rocker arm assembly to 26 N·m (226 in.-lbs.) torque and adjust valve clearance as outlined in paragraph 57. Tighten rocker arm cover retaining nuts to 18 N·m (160 in.-lbs.) torque.

## CAM FOLLOWERS

### 855, 856 and 955 Models

**61.** Cam followers can be removed from 3TN75RJ and 3TN84RJ engine models after removing rocker arms assembly, push rods and engine camshaft as outlined in paragraph 69. Invert engine before removing camshaft, remove camshaft, then lift cam followers from block bores. Cam followers must be identified as they are removed so that they can be reinstalled in same location. Minimum outside diameter of cam follower is 11.93 mm (0.470 in.). Maximum inside diameter of cam follower bore in cylinder block is 12.05 mm (0.474 in.). Cam follower to bore clearance should be 0.010-0.043 mm (0.0003-0.0016 in.), but should not exceed 0.10 mm (0.004 in.).

## TIMING GEARS

### 855, 856 And 955 Models

**62.** The timing gear train consists of crankshaft gear (1—Fig. 30), idler gear (2), camshaft gear (3) and fuel injection pump drive gear (4). Timing gear marks are visible after removing the timing gear cover as outlined in paragraph 67. Valves and injection pump are properly timed to crankshaft when marks on gears are aligned as shown in Fig. 30. Because of the number of teeth on idler gear, timing marks will not align every other revolution of the crankshaft. Alignment can be checked by turning crankshaft until keyway is vertical, then checking location of mark on camshaft gear. If mark on camshaft gear is away from idler gear, turn crankshaft one complete turn until keyway is again vertical. Remove idler gear, then reinstall idler gear with marks on all gears aligned as shown in Fig. 30.

NOTE: Before removing any gears, first refer to paragraph 60 and remove rocker arm assembly to avoid possible damage to piston or valve train. Damage could result if either camshaft or crankshaft is turned independently from the other unless rocker arms are removed.

Backlash between timing gears (1, 2, 3 or 4) and any one of the other timing gears should be less than 0.20 mm (0.008 in.). Backlash between oil pump gear

*Fig. 30—View of timing gears typical of 855, 856 and 955 models.*

(5) and crankshaft gear (1) should also be less than 0.20 mm (0.008 in.).

All timing gears (1, 2, 3 and 4) have a single timing mark at (T1, T2 and T3). Marks on idler gear (2) are located on teeth and marks on timing gears (1, 3 and 4) are located in tooth valley.

**63. IDLER GEAR AND SHAFT.** To inspect or remove the idler gear (2—Fig. 30) and shaft (12—Fig. 31), first remove timing gear cover as outlined in paragraph 67. Check backlash between idler gear and other timing gears before removing the gear. Backlash between idler gear and any other gear (1, 3 or 4—Fig. 30) should be less than 0.20 mm (0.008 in.).

NOTE: Before removing any gears, first refer to paragraph 60 and remove rocker arm assembly to avoid possible damage to piston or valve train. Damage could result if either camshaft or crankshaft is turned independently from the other unless rocker arms are removed.

Remove the two screws (15—Fig. 31) retaining idler shaft to front of cylinder block, then withdraw gear (2) and idler shaft (12). Inside diameter of idler gear bushing should be 46.0-46.025 mm (1.811-1.812 in.), but should not exceed 46.08 mm (1.814 in.). Outside diameter of shaft should be 45.950-45.975 mm (1.809-1.810 in.), but should not be less than 45.93 mm (1.808 in.). Renew shaft and/or bushing inside gear if clearance exceeds 0.15 mm (0.006 in.).

Oil hole in bushing must be aligned with hole in idler gear. Inspect gear teeth for wear or scoring.

To install idler gear, position crankshaft with rear (number 1) piston at top dead center. Turn camshaft gear (3—Fig. 30) and fuel injection pump drive gear (4) so that timing marks point to center of idler gear shaft. Install idler gear (2) and shaft (12—Fig. 31) so that all timing marks are aligned as shown in Fig. 30, then install idler shaft retaining screws. Tighten retaining screws to 26 N·m (226 in.-lbs.) torque.

**64. CAMSHAFT GEAR.** The camshaft must be removed as outlined in paragraph 69 before gear (3—Fig. 31) can be pressed from shaft. Check camshaft end play and backlash between camshaft gear (3—Fig. 30) and idler gear (2) before removing camshaft. Backlash should be 0.04-0.12 mm (0.0016-0.0047 in.), but should not exceed 0.20 mm (0.008 in.). Camshaft end play should be 0.05-0.20 mm (0.002-0.008 in.), but should not exceed 0.4 mm (0.016 in.). If end play is excessive, install new thrust plate (20—Fig. 31).

Remove rocker arm assembly as outlined in paragraph 60 and idler gear as outlined in paragraph 63. Damage could result if either camshaft or crankshaft is turned independently from the other unless rocker arms are removed. Remove camshaft as outlined in paragraph 69. Gear can be pressed from camshaft to renew thrust plate (20) or gear (3).

When reinstalling camshaft in all models, tighten screws retaining thrust plate to front of engine to 26 N·m (226 in.-lbs.) torque. Refer to paragraph 60 for installation of rocker arm assembly, paragraph 67 for installation of timing gear cover and to paragraph 57 for adjusting valve clearance.

**65. CRANKSHAFT GEAR.** Crankshaft gear (1—Fig. 30) should not be removed unless a new gear is to be installed. Backlash between crankshaft gear (1) and idler gear (2) should be 0.11-0.19 mm (0.0043-0.0075 in.), but should not exceed 0.20 mm (0.008 in.). Backlash between crankshaft gear (1) and oil pump gear (5) should be within the same limits.

To remove crankshaft gear (1), the crankshaft should be removed and gear pressed from shaft.

**66. INJECTION PUMP DRIVE GEAR.** Refer to paragraph 87 for removal or other service procedures to the injection pump drive gear (4—Fig. 31).

## TIMING GEAR COVER AND CRANKSHAFT FRONT OIL SEAL

### 855, 856 And 955 Models

**67.** Crankshaft front oil seal (23—Fig. 31) can be removed and new seal installed after removing crankshaft pulley. Seal should be driven into cover

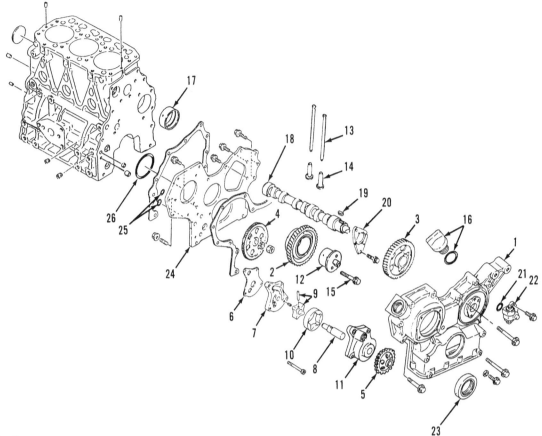

**Fig. 31—Exploded view of timing gear cover and timing gear housing typical of 855, 856 and 955 models.**

1. Timing gear cover
2. Idler gear & bushing
3. Camshaft drive gear
4. Injection pump drive gear
5. Oil pump drive gear
6. Gasket
7. Oil pump cover
8. Pump shaft
9. Inner rotor & drive pin
10. Outer rotor
11. Pump housing
12. Idler shaft
13. Push rod
14. Cam follower
15. Retaining screws
16. Filler cap & gasket
17. Camshaft bearing
18. Valve camshaft
19. Drive key
20. Thrust plate
21. "O" ring
22. Tachometer gear unit
23. Crankshaft front seal
24. Timing gear housing
25. "O" rings
26. "O" ring

until flush with front of cover. Lubricate seal before installing crankshaft pulley.

To remove timing gear cover (1), first refer to paragraph 54 and remove engine. Remove fan and alternator. Remove retaining screw from center of crankshaft pulley. Note that screw in center of crankshaft pulley should have "Loctite" or equivalent on threads and may be difficult to loosen. Use an appropriate puller and remove crankshaft pulley from crankshaft. Unbolt and remove timing gear cover fom front of engine.

Clean mating surfaces of timing gear cover and timing gear housing, then apply an even bead of sealer to mating surface. A gasket is not used between cover and housing. Install cover and tighten retaining screws to 26 N·m (226 in.-lbs.) torque. The crankshaft pulley of some models is equipped with a pin which must engage a hole in crankshaft gear. Coat threads of center screw with medium strength "Loctite" or equivalent and tighten to 115 N·m (85 ft.-lbs.) torque. Remainder of assembly is reverse of disassembly.

## TIMING GEAR HOUSING

### 855, 856 And 955 Models

**68.** To remove timing gear housing (24—Fig. 31), first remove engine as outlined in paragraph 54, the rocker arm cover and rocker arm assembly as outlined in paragraph 60, timing gear cover (1) as outlined in paragraph 67, camshaft and drive gear (3 and 18) as outlined in paragraph 69, and injection pump drive gear (4) as outlined in paragraph 87. Unbolt and remove oil pump (5 through 11) from timing gear housing and oil pan (sump) from bottom of engine, then unbolt timing gear housing (24) from front of cylinder block.

When reinstalling timing gear housing, reverse removal procedure. Clean and dry sealing surfaces of timing gear housing and cylinder block, then apply bead of RTV sealer on block mating surface of timing gear housing. Install timing gear housing and tighten retaining screws to 20 N·m (177 in.-lbs.) torque.

## CAMSHAFT

### 855, 856 And 955 Models

**69.** To remove the engine camshaft, first remove rocker arm cover and rocker arm assembly as outlined in paragraph 60 and timing gear cover as outlined in paragraph 67. Check camshaft end play and backlash between camshaft gear (3—Fig. 31) and idler gear (2) before removing camshaft. Backlash should be 0.04-0.12 mm (0.0016-0.0047 in.), but should not exceed 0.20 mm (0.008 in.). Camshaft end play should be 0.05-0.20 mm (0.002-0.008 in.), but should not exceed 0.4 mm (0.16 in.). If end play is excessive, install new thrust plate (20). Remove and invert engine to hold cam followers (14) away from camshaft (18).

> NOTE: It may not be necessary to turn engine up-side-down if magnets are available to hold all of the cam followers away from camshaft.

Remove camshaft thrust plate retaining screws through the web holes of camshaft gear (3), then pull camshaft (18) out of engine block bores. Gear can be pressed from camshaft to renew thrust plate (20) or gear. Flywheel must be removed to remove plug from rear of camshaft bore in cylinder block. Refer to the following specifications.

### 855, 856 and 955 Models

| | |
|---|---|
| Engine models | 3TN75RJ and 3TN84RJ |
| Camshaft journal OD — | |
| Front | 44.925-44.950 mm |
| | (1.769-1.770 in.) |
| Wear limit | 44.80 mm |
| | (1.764 in.) |
| Intermediate | 44.910-44.935 mm |
| | (1.768-1.769 in.) |
| Wear limit | 44.80 mm |
| | (1.764 in.) |
| Rear | 44.925-44.950 mm |
| | (1.769-1.770 in.) |
| Wear limit | 44.80 mm |
| | (1.764 in.) |
| Camshaft front bearing ID | 44.990-45.055 mm |
| | (1.771-1.774 in.) |
| Wear limit | 45.10 mm |
| | (1.776 in.) |
| Camshaft end play | 0.05-0.20 mm |
| | (0.002-0.008 in.) |
| Wear limit | 0.4 mm |
| | (0.16 in.) |
| Cam lobe height | 38.635-38.765 mm |
| | (1.521-1.526 in.) |
| Wear limit | 38.40 mm |
| | (1.512 in.) |

Use 1¾ inch piloted driver to install new bushing (17) and be sure to align oil hole in bushing with

passage in cylinder block. Cam gear (3) should be heated to approximately 150° C (300° F) before pressing onto camshaft. Be sure that thrust plate (20) is installed and will turn freely after gear is pressed firmly against shoulder of camshaft.

When reinstalling camshaft in all models, tighten screws retaining thrust plate to front of engine to 26 N·m (226 in.-lbs.) torque. Refer to paragraph 60 for installation of rocker arm assembly, paragraph 67 for installation of timing gear cover and to paragraph 57 for adjusting valve clearance.

## ROD AND PISTON UNITS

### 855, 856 And 955 Models

**70.** Piston and connecting rod units are removed from above after removing oil pan and cylinder head.

Drain engine oil and coolant, then refer to paragraph 54 and remove the engine assembly. Remove the cylinder head as outlined in paragraph 56 and oil pan. Be sure that ridge (if present) is removed from top of cylinder bore, then make sure piston, rod and cap are marked before removing pistons. If not already marked, stamp cylinder number on piston, rod and cap before removing. Number "1" cylinder is at rear and connecting rods should be marked on injection pump (right) side. Keep rod bearing cap with matching connecting rod.

Piston must be assembled to connecting rod with recess in top of piston opposite slot in connecting rod for bearing insert tang. Screws attaching connecting rod cap to rod should not be reused. Always install **new** connecting rod screws.

Lubricate cylinder, crankpin and ring compressor with clean engine oil. Space piston ring end gaps evenly around piston, but not aligned with piston pin, then compress piston rings with suitable compressor tool. Install piston and rod units in cylinder block with recess in top of piston on same side as fuel injection pump (right). Make sure that bore in connecting rod and cap is completely clean and install bearing inserts with tangs engaging slot in rod and cap. Make sure that inserts are firmly seated, lubricate crankpin and bearing inserts, then seat connecting rod with insert against crankpin. Install connecting rod cap with insert firmly seated and lubricated. Tangs on both bearing inserts should be on left (camshaft) side of engine. Tighten connecting rod cap retaining screws to 39 N·m (29 ft.-lbs.) torque on 855 and 856 models; 47 N·m (35 ft.-lbs.) torque for 955 models.

Clearance between flat top of piston and cylinder head can be measured using 10 mm (0.4 in.) long sections of 1.5 mm (0.06 in.) diameter soft lead wire positioned on top of piston. Turn crankshaft one complete revolution, then remove lead wire and measure thickness of flattened wire. Clearance should be 0.59-

0.77 mm (0.023-0.030 in.) for 855 and 856 models; 0.64-0.82 mm (0.025-0.032 in.) for 955 models.

Clean mating surfaces of crankcase extension and the oil pan thoroughly. Apply a bead of "Form-A-Gasket" or equivalent to sealing surfaces and install oil pan retaining screws to 27 N·m (20 ft.-lbs.) torque. Remainder of assembly is reverse of disassembly.

## PISTON, RINGS AND CYLINDER

### 855, 856 And 955 Models

**71.** Pistons are fitted with two compression rings and one oil control ring. Piston pin is full floating and retained by a snap ring at each end of pin bore in piston. If necessary to separate piston from connecting rod, remove snap rings and push pin from piston and connecting rod. Service pistons are available in standard size and 0.25 mm (0.010 in.) oversize.

Inspect all piston skirts for scoring, cracks and excessive wear. On 855 and 856 models, measure piston skirt at right angles to pin bore 12.5 mm (0.492 in.) from bottom of skirt. Minimum skirt measurement for standard size piston is 74.81 mm (2.945 in.).

On 955 models, measure piston skirt at right angles to pin bore 29.0 mm (1.142 in.) from bottom of skirt. Minimum skirt measurement for standard size piston is 83.80 mm (3.299 in.).

Piston pin bore in piston should not exceed 23.02 mm (0.906 in.) for 855 and 856 models; 26.02 mm (1.024 in.) for 955 models. Clearance between piston pin and bore in piston should not exceed 0.05 mm (0.002 in.).

Standard cylinder bore diameter for 855 and 856 models should be 75.00-75.03 mm (2.953-2.954 in.) and should not exceed 75.20 mm (2.961 in.). Piston to cylinder bore clearance should not exceed 0.22 mm (0.009 in.). Cylinders can be rebored and fitted with 0.25 mm (0.10 in.) oversize pistons.

Standard cylinder bore diameter for 955 models should be 84.00-84.03 mm (3.307-3.308 in.) and should not exceed 84.20 mm (3.315 in.). Piston to cylinder bore clearance should not exceed 0.30 mm (0.012 in.). Cylinders can be rebored and fitted with 0.25 mm (0.10 in.) oversize pistons.

Cylinder bore of all models should be deglazed using 180 grit stone and crosshatch pattern should intersect at 30-40°. Wash abrasive from cylinder walls using warm soapy water. Continue cleaning until white cleaning cloths show no discoloration. Make sure all soap is rinsed from cylinders, dry cylinder block thoroughly, then lubricate with clean engine oil.

NOTE: Use only mild soap to clean abrasive from cylinders. Do not use commercial solvents, gasoline or kerosene to clean cylinders.

End gap of piston rings in cylinder should not exceed 1.50 mm (0.059 in.). Rings should be inserted squarely in cylinder using piston to push ring approximately 30 mm (1.181 in.) into cylinder.

Piston ring side clearance in grooves should not exceed 0.25 mm (0.010 in.) for top and second (compression) rings or 0.20 mm (0.008 in.) for oil control (bottom) ring. Renew piston if ring side clearance is excessive.

When installing piston rings, install oil ring expander in groove, followed by the oil ring. End gap of oil ring should be 180° from ends of expander. Install compression ring with inside chamfer or identification mark toward top of piston. Space end gaps of top, second and oil control rings 120° apart before installing in cylinder bore.

## PISTON PIN

### 855, 856 And 955 Models

**72.** Piston pin bore in piston should not exceed 23.02 mm (0.906 in.) for 855 and 856 models; 26.02 mm (1.024 in.) for 955 models. Clearance between piston pin and bore in piston should not exceed 0.05 mm (0.002 in.).

Piston pin diameter should be measured at each end and in center. Measure pin diameter at each of the three locations and again 90° from each of the original locations. Piston pin diameter for 855 and 856 models should not be less than 22.90 mm (0.902 in.) and pin diameter should not be less than 25.90 mm (1.020 in.) for 955 models.

Inside diameter of pin bushing in top of connecting rod should not exceed 23.10 mm (0.909 in.) for 855 and 856 models. For 955 models, pin bushing should not exceed 26.10 mm (1.028 in.). Clearance between piston pin and bore of pin bushing in connecting rod should not exceed 0.11 mm (0.004 in.).

## CONNECTING ROD AND BEARINGS

### 855, 856 And 955 Models

**73.** Connecting rod can be unbolted from crankshaft crankpin and new bearing inserts installed from below after removing oil pan. Refer to paragraph 70 for removing connecting rods. Connecting rod should not be out-of-parallel or twisted more than 0.08 mm (0.003 in.).

Inside diameter of pin bushing in top of connecting rod should be 23.025-23.038 mm (0.906-0.907 in.), but should not exceed 23.10 mm (0.909 in.) for 855 and 856 models. For 955 models, pin bushing should be 26.025-26.038 mm (1.0246-1.0251 in.), but should not exceed 26.10 mm (1.028 in.). For all models, clearance between piston pin and bushing bore in connecting rod should be 0.025-0.047 mm (0.0010-0.0019 in.), but should not exceed 0.11 mm (0.004 in.).

Connecting rod bearing clearance on crankshaft crankpin should be 0.038-0.090 mm (0.0015-0.0035 in.) and wear limit should not exceed 0.16 mm (0.006 in.). Connecting rod side play on crankpin should be 0.2-0.4 mm (0.0079-0.0157 in.), but should not exceed 0.55 mm (0.022 in.).

Crankpin standard diameter for 855 and 856 models is 42.952-42.962 mm (1.691-1.6914 in.). Crankpin standard diameter for 955 models is 47.952-47.962 mm (1.8879-1.8883 in.). Undersized bearing inserts, which require crankshaft journals to be resized and polished, are available for service. Finished undersize journals should provide standard clearance with undersized bearings.

Refer to paragraph 70 if connecting rod and piston units are removed from engine. Make sure that bore in connecting rod and cap is completely clean and install bearing inserts with tangs engaging slot in rod and cap. Make sure that inserts are firmly seated, lubricate crankpin and bearing inserts, then seat connecting rod with insert against crankpin. Install connecting rod cap with insert firmly seated and

lubricated. Tangs on both halves of bearing should be on left (camshaft) side of engine. Tighten connecting rod cap retaining screws to 39 N·m (29 ft.-lbs.) torque on 855 and 856 models. Tighten connecting rod cap retaining screws to 47 N·m (35 ft.-lbs.) torque. Clean and dry sealing surface of oil pan and crankshaft extension, then coat sealing surface with RTV sealer and install oil pan. Apply a bead of "Form-A-Gasket" or equivalent to sealing surfaces and install oil pan retaining screws to 27 N·m (20 ft.-lbs.) torque. Remainder of assembly is reverse of disassembly.

## CRANKSHAFT AND MAIN BEARINGS

### 855, 856 And 955 Models

**74.** The crankshaft is supported in four main bearings. Crankshaft end play is controlled by insert type thrust bearing (8 and 10—Fig. 32) on sides of rear (number 1) main bearing journal.

To remove crankshaft (5), remove engine from tractor as outlined in paragraph 54, the rocker arm cover and rocker arm assembly as outlined in paragraph

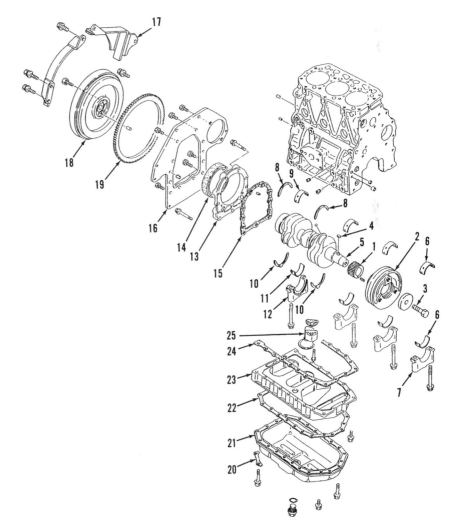

Fig. 32—View of crankshaft, rear seal and flywheel typical of 855, 856 and 955 models.

1. Crankshaft gear
2. Crankshaft pulley
3. Retaining screw
4. Drive key
5. Crankshaft
6. Bearing insert set
7. Main bearing cap
8. Thrust bearing halves (top)
9. Rear main bearing insert (top)
10. Thrust bearing halves (bottom)
11. Rear main bearing insert (bottom)
12. Rear main bearing cap
13. Crankshaft rear seal retainer housing
14. Rear seal
15. Gasket
16. Engine rear plate
17. Flywheel shield
18. Flywheel
19. Ring gear
20. Lock plates
21. Oil pan
22. Gasket
23. Crankcase extension
24. Gasket
25. Oil pickup

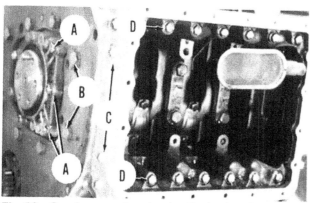

*Fig. 33—Crankcase extension is attached to the cylinder block (crankcase) with screws (C and D). Extension is attached to the engine rear plate and rear seal housing with screws (A and B).*

60, the timing gear cover as outlined in paragraph 67, camshaft as outlined in paragraph 69, injection pump drive gear and camshaft as outlined in paragraph 87 and timing gear housing as outlined in paragraph 68. Remove flywheel (18), then remove crankshaft rear oil seal as outlined in paragraph 76. The three lower screws (A—Fig. 33) retaining the oil seal retainer housing and the two screws (B) directly below seal retainer housing are threaded into the crankcase extension housing. Remove screws (C and D) attaching the crankcase extension housing to the cylinder block (crankcase) and remove the extension housing.

Unbolt and remove the engine rear plate (16—Fig. 32). Invert engine, then refer to paragraph 70 and detach connecting rods from crankshaft. Mark all main bearing caps for correct installation, unbolt and remove crankshaft main bearing caps, then lift crankshaft from engine. The rear main bearing is number "1."

Check all main bearing journals, thrust surfaces of rear journal and connecting rod crankpin journals for scoring or excessive wear. Main bearings and rod bearings are available in standard size and 0.25 mm (0.010 in.) undersize. Refer to the following specifications.

### 855 and 856
Engine model . . . . . . . . . . . . . . . . . . . . . . . 3TN75RJ
Crankpin standard OD —
   Desired . . . . . . . . . . . . . . . . . 42.952-42.962 mm
                                    (1.691-1.6914 in.)
   Wear limit . . . . . . . . . . . . . . . . . . . . 42.91 mm
                                      (1.689 in.)
Crankpin to rod bearing clearance —
   Desired . . . . . . . . . . . . . . . . . . 0.038-0.090 mm
                                  (0.0015-0.0035 in.)
   Wear limit . . . . . . . . . . . . . . . . . . . . . 0.16 mm
                                      (0.006 in.)

Main journal standard OD —
   Desired . . . . . . . . . . . . . . . . . 46.952-46.962 mm
                                      (1.8485-1.8489 in.)
   Wear limit . . . . . . . . . . . . . . . . . . . . 46.91 mm
                                      (1.847 in.)

Main journal to bearing clearance —
   Desired . . . . . . . . . . . . . . . . . . 0.038-0.093 mm
                                    (0.0015-0.0037 in.)
   Wear limit . . . . . . . . . . . . . . . . . . . . . 0.15 mm
                                      (0.0059 in.)

Crankshaft end play —
   Desired . . . . . . . . . . . . . . . . . . . 0.09-0.271 mm
                                    (0.004-0.011 in.)
   Wear limit . . . . . . . . . . . . . . . . . . . . . 0.33 mm
                                      (0.013 in.)
Main cap torque . . . . . . . . . . . . . . . . . . . . . 54 N•m
                                      (40 ft.-lbs.)

Rear oil seal case torque —
   Seal to block . . . . . . . . . . . . . . . . . . . . 26 N•m
                                      (266 in.-lbs.)
   Oil pan to seal case . . . . . . . . . . . . . . . . 9 N•m
                                      (78 in.-lbs.)

### 955
Engine model . . . . . . . . . . . . . . . . . . . . . . . 3TN84RJ
Crankpin standard OD —
   Desired . . . . . . . . . . . . . . . . . 47.952-47.962 mm
                                      (1.8879-1.8883 in.)
   Wear limit . . . . . . . . . . . . . . . . . . . . 47.91 mm
                                      (1.886 in.)
Crankpin to rod bearing clearance —
   Desired . . . . . . . . . . . . . . . . . . 0.038-0.090 mm
                                    (0.0015-0.0035 in.)
   Wear limit . . . . . . . . . . . . . . . . . . . . . 0.16 mm
                                      (0.006 in.)
Main journal standard OD —
   Desired . . . . . . . . . . . . . . . . . 49.952-49.962 mm
                                      (1.9666-1.9670 in.)
   Wear limit . . . . . . . . . . . . . . . . . . . . 49.90 mm
                                      (1.965 in.)
Main journal to bearing clearance —
   Desired . . . . . . . . . . . . . . . . . . 0.038-0.093 mm
                                    (0.0015-0.0037 in.)
   Wear limit . . . . . . . . . . . . . . . . . . . . . 0.15 mm
                                      (0.006 in.)
Crankshaft end play —
   Desired . . . . . . . . . . . . . . . . . . 0.090-0.271 mm
                                    (0.004-0.011 in.)
   Wear limit . . . . . . . . . . . . . . . . . . . . . 0.33 mm
                                      (0.013 in.)
Main cap torque . . . . . . . . . . . . . . . . . . . . . 98 N•m
                                      (72 ft.-lbs.)
Rear oil seal case torque —
   Seal to block . . . . . . . . . . . . . . . . . . . . 26 N•m
                                      (266 in.-lbs.)
   Oil pan to seal case . . . . . . . . . . . . . . . . 9 N•m
                                      (78 in.-lbs.)

If crankshaft gear (1—Fig. 32) was removed, heat gear to approximately 150° C (300° F) before pressing gear onto crankshaft. Be sure that all bearing surfaces are clean and free of nicks or burrs. Half of bearing insert which has oil hole must be installed in cylinder block and bearing shell half without oil hole should be installed in main bearing cap. Bearing sets for all journals are alike. Lubricate main bearings, then set clean crankshaft in bearings. Lubricate thrust bearing upper halves (without projections) and insert bearing halves between block and crankshaft with oil groove side toward crankshaft shoulders. Use light grease to stick thrust washer lower halves (10) to the sides of rear main bearing cap (12) with tabs aligned with notches in cap. Install bearing insert (11) in cap, lubricate bearing and install cap in original location. Arrow on side of main bearing cap marked "FW" points to flywheel end of engine. Lubricate threads of main bearing cap retaining screws with engine oil and tighten to torque listed in specifications. Check to be sure that crankshaft turns freely before completing assembly.

When installing the crankcase extension, observe the following. Clean mating surfaces of extension, engine rear plate, timing gear housing (front plate), the timing gear cover and the oil pan thoroughly. Apply a bead of "Form-A-Gasket" or equivalent to sealing surfaces and install the extension housing. Tighten screws (C and D—Fig. 33) attaching the crankcase extension housing to the cylinder block (crankcase) to 27 N•m (20 ft.-lbs.) torque. Tighten screws (B) to 49 N•m (36 ft.-lbs.) torque and screws (A) to 27 N•m (20 ft.-lbs.) torque. Tighten screws retaining oil pan to crankcase extension to 27 N•m (20 ft.-lbs.) torque and screws attaching timing gear housing to crankcase extension to 22 N•m (16 ft.-lbs.) torque. Remainder of assembly is reverse of disassembly.

## CRANKSHAFT FRONT OIL SEAL

### 855, 856 And 955 Models

**75.** The crankshaft front oil seal (23—Fig. 31) is located in timing gear cover and can be renewed after removing crankshaft pulley (2—Fig. 32). Pry seal from timing cover, taking care not to damage cover. Seal lip rides on hub of crankshaft pulley. Inspect pulley hub for excessive wear at point of seal contact. Install new seal with lip toward inside using suitable driver. Lubricate seal lip and crankshaft pulley hub, install pulley on crankshaft and tighten pulley retaining screw. The crankshaft pulley of some models is equipped with a pin which must engage a hole in crankshaft gear. Coat threads of center screw (3) with medium strength "Loctite" or equivalent and tighten to 115 N•m (85 ft.-lbs.) torque.

## CRANKSHAFT REAR OIL SEAL

### 855, 856 And 955 Models

**76.** The crankshaft rear oil seal (14—Fig. 32) can be renewed after removing flywheel and seal retainer housing (13). Install new oil seal with lip toward inside. Press seal into housing until flush with outer surface of retainer housing. If crankshaft is grooved at original oil seal contact surface, new seal can be installed 3 mm (⅛ in.) farther into seal housing. Lubricate seal lip and install retainer housing. If not equipped with gasket (15), coat sealing surface of retainer housing with RTV sealer. Tighten screws attaching oil seal retainer housing to rear of cylinder block to 26 N•m (226 in.-lbs.) and screws attaching oil pan to bottom of retainer housing to 9 N•m (78 in.-lbs.) torque.

## FLYWHEEL

### 855, 856 And 955 Models

**77.** To remove flywheel, first remove engine as outlined in paragraph 54. Unbolt and remove flywheel shield (17—Fig. 32), then unbolt and remove flywheel from rear of crankshaft. The engine rear plate (16) can be unbolted and removed from rear of engine block after removing flywheel.

When assembling, tighten screws attaching engine rear plate (16) to rear of engine to 49 N•m (36 ft.-lbs.) torque and screws attaching flywheel (18) to crankshaft to 83 N•m (61 ft.-lbs.) torque. Install flywheel shield (17) and tighten retaining screws or nuts to the following torque values: M8 to 26 N•m (226 in.-lbs.); M10 to 49 N•m (36 ft.-lbs.); M12 (nut) to 88 N•m (65 ft.-lbs.).

## OIL PAN

### 855, 856 And 955 Models

**78.** Drain engine oil, then refer to paragraph 54 and remove engine. Unbolt and remove oil pan from engine. Oil pickup (25—Fig. 32) can be removed for cleaning.

Clean and dry sealing surfaces for oil pickup tube and oil pan, then install new gasket or coat sealing surface with RTV sealer. Tighten screws retaining oil pickup and oil pan to 26 N•m (226 in.-lbs.) torque.

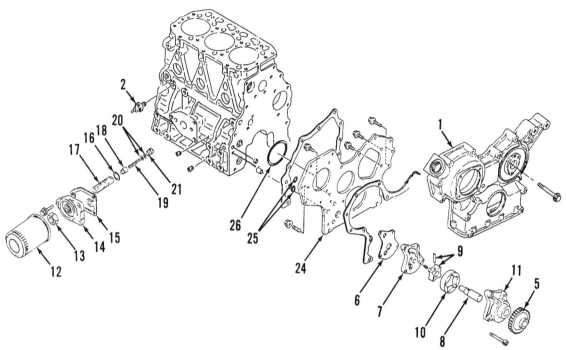

*Fig. 34—View of oil pump, filter and relief valve used on 855 and 856 models. Oil line to injection pump, similar to (3--Fig. 35), is not shown.*

1. Timing gear cover
2. Pressure gauge sensor
5. Pump drive gear
6. Gasket
7. Oil pump cover
8. Pump shaft
9. Inner rotor & drive pin
10. Outer rotor
11. Pump housing
12. Oil filter
13. Nut
14. Adapter housing
15. Gasket
16. "O" ring
17. Adapter
18. Relief valve poppet
19. Spring
20. Adjusting shims
21. Valve retainer
24. Timing gear housing
25. "O" rings
26. "O" ring

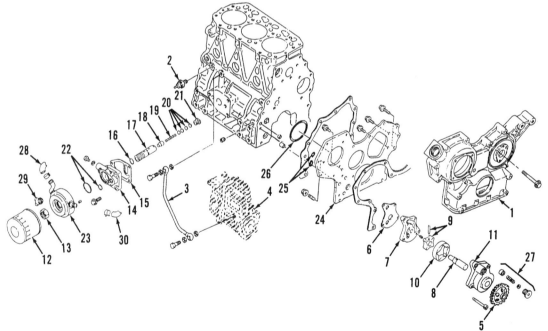

*Fig. 35—View of oil pump, oil cooler, filter and relief valve used on 955 models.*

1. Timing gear cover
2. Pressure gauge sensor
3. Oil line to injection pump
4. Fuel injection pump
5. Pump drive gear
6. Gasket
7. Oil pump cover
8. Pump shaft
9. Inner rotor & drive pin
10. Outer rotor
11. Pump housing
12. Oil filter
13. Nut
14. Adapter housing
15. Gasket
16. "O" ring
17. Adapter
18. Relief valve poppet
19. Spring
20. Adjusting shims
21. Valve retainer
22. "O" rings
23. Oil cooler
24. Timing gear housing
25. "O" rings
26. "O" ring
27. Surge relief valve
28. Pump from block
29. Cooler drain
30. Hose to coolant pump

## OIL PUMP AND PRESSURE RELIEF VALVE

### 855, 856 And 955 Models

**79.** The gerotor type oil pump is mounted on front of engine timing gear housing . Drive gear (5—Fig. 34 or Fig. 35) on pump shaft is driven by crankshaft gear.

To remove oil pump, first remove timing gear cover as outlined in paragraph 67, then measure backlash between oil pump gear (5) and crankshaft gear. For 855 and 856 models, backlash should be 0.11-0.19 mm (0.0043-0.0075 in.), but should not exceed 0.25 mm (0.010 in.). For 955 models, backlash should be 0.04-0.12 mm (0.0016-0.0047 in.), but should not exceed 0.20 mm (0.008 in.). Unbolt and remove pump.

Pump end cover (7—Fig. 34 and Fig. 35) can be removed from pump body (11) to inspect pump. End clearance of rotors in pump body should not exceed 0.13 mm (0.005 in.). Clearance between outer rotor (10) and pump housing should not exceed 0.25 mm (0.10 in.). Clearance between tip of inner rotor (9) and center of outer rotor lobe should not exceed 0.25 mm (0.10 in.) for 855 and 856 models or 0.15 mm (0.006 in.) for 955 models. On all models, pump drive gear (5) can be pressed from shaft (8) if renewal is re-quired, but service pump is available with drive gear. Mark on side of outer rotor (10) should be toward inside of pump housing bore.

Install pump using new gasket and tighten retaining screws to 25 N·m (18 ft.-lbs.) torque. Refer to paragraph 67 for installing timing gear cover.

The oil pressure relief valve (18 through 21—Fig. 34 or Fig. 35) is located inside threaded oil filter adapter. Relief valve pressure can be adjusted by removing valve retainer (21) and adding shims (20). Adding or removing one 1 mm (0.039 in.) thick shim will change oil pressure about 15.6 kPa (2.3 psi). Free length of spring (19) is 46.0 mm (1.81 in.) and spring should exert 20.5 N (4.6 lbs.) when compressed to 27.5 mm (1.08 in.).

On 955 models, the oil pressure surge relief valve (27—Fig. 35) is located on oil pump.

On all models, relief valve poppet (18—Fig. 34 or Fig. 35) should move freely in bore of body (17) and should not stick. Screws attaching adapter housing (14) to engine should be tightened to 27 N·m (20 ft.-lbs.) torque. If nut (13—Fig. 34) is loosened on 855 or 856 models, tighten to 30 N·m (22 ft.-lbs.) torque. Nut (13—Fig. 35) retaining oil cooler to adapter housing should be tightened to 27-34 N·m (20-25 ft.-lbs.) torque.

# DIESEL FUEL SYSTEM

**80.** Models 655, 755 and 756 are equipped with a three piston diesel injection pump which is driven by a separate fuel injection camshaft located in the timing gear housing. The pump is controlled by the governor attached to timing gear housing at rear of injection pump camshaft. Nozzles inject fuel into precombustion chambers located in cylinder head.

Models 855, 856 and 955 are equipped with a three piston diesel injection pump attached to the timing gear housing on right side of engine. Pump governor is integral with pump assembly.

### FUEL FILTERS AND LINES

#### All Models

**81. OPERATION AND MAINTENANCE.** The use of good quality, approved fuel, the careful storage and proper handling of fuel for the diesel engine can not be over emphasized.

Because of extremely close tolerances and precise requirements of all diesel components, it is of utmost importance that clean fuel and careful maintenance be practiced at all times. Unless service personnel have been trained to adjust, disassemble and repair the specific injection system and necessary special tools are available, limit service to pump and nozzles to removal, installation and exchange of complete assemblies. It is impossible to recalibrate an injection pump or reset an injector without proper specifications, equipment and training.

A renewable paper filter element type fuel filter is located between fuel feed pump and fuel injection pump. The fuel filter housing incorporates a fuel shut-off valve on some models, but not others. Check the clear filter sediment bowl daily for evidence of water or sediment. Bowl should be drained and new filter installed at least every 200 hours of operation or if filter appears dirty. Air must be bled from filter and fuel system as outlined in paragraph 82 each time filter is removed.

The fuel system should also be bled if the fuel tank has been allowed to run dry, if components within the system have been disconnected or removed, or if the engine has not been operated for a long time. If the engine starts, then stops, the cause could be air in the system. Refer to paragraph 82 for bleeding procedure.

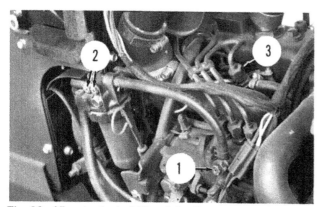

*Fig. 36—View typical of filter, injection pump and nozzles installed on 655, 755 and 756 models. Models 855, 856 and 955 are equipped with a fuel shut-off valve integral with filter. Refer to text for bleeding procedure.*

**82. BLEEDING.** Before attempting to bleed the fuel system, make sure that sufficient supply of clean fuel is contained in the tractor fuel tank.

On 655, 755 and 756 models, loosen both bleed screws (2—Fig. 36) and turn key switch ON (first position). The fuel transfer electric pump should begin to operate and pump fuel from opened bleeder screws. Close bleeder screws when fuel without air bubbles flows from valves. Loosen bleed screw (1) on injection pump and allow fuel with bubbles to flow from opened bleeder screw. Close pump bleed screw when fuel flowing from opened screw is free of bubbles.

On 855, 856 and 955 models, loosen both bleed screws (2—Fig. 36) and open fuel shut-off valve located on filter housing. Fuel should begin to flow from opened bleeder screws. If equipped with an electric fuel pump, it will be necessary to turn key switch ON (first position) and fuel pump should pump fuel from opened bleed screws. If equipped with mechanical fuel transfer pump on side of injection pump, it will be necessary to crank engine with starter to pump fuel. Close bleeder screws when fuel without air bubbles runs from the opened bleeder screws. Loosen bleed screw (1) on injection pump and allow fuel with bubbles to flow from opened bleeder screw. Close pump bleed screw when fuel flowing from opened screw is free of bubbles.

On all models, engine will usually start after bleeding at filters and pump, but it may be necessary to advance throttle slightly, loosen fittings (3) at each injection nozzle, then crank engine with the starter to remove air from high pressure lines and nozzles. Tighten compression fittings when fuel without air begins to flow from loosened connection.

## INJECTION PUMP

**83.** Only trained service personnel with proper tools and equipment should attempt to disassemble

or adjust the fuel injection pump. Specific procedures must be performed under conditions of strict cleanliness and exact measurements require accurate test equipment.

## 655, 755 And 756 Models

**84. TIMING TO ENGINE.** Timing of the injection pump camshaft and drive gear (4—Fig. 21) is considered correct if marks (T1 and T2) are aligned as shown. Specific injection timing can be changed by increasing or decreasing thickness of shims (37—Fig. 37). The old shims should not be reinstalled. Always install pump using new shims of same thickness as shims that were originally installed. Thickness of shims should not be changed except by specially trained personnel. If original thickness is not known, manufacturer suggests that one new 0.5 mm (0.020 in.) thick shim be installed. Make sure that old paint and sealer are cleaned from mating surface of pump and timing gear housing. Refer to paragraph 85 for removal of pump.

**85. REMOVE AND REINSTALL.** To remove the injection pump, clean pump and area around pump and lines. Disconnect fuel inlet line to pump, loosen fittings at both ends of high pressure discharge lines between pump and nozzles, then remove high pressure lines. Cover all openings in pump, nozzles and fuel lines to prevent the entrance of dirt. Detach solenoid linkage by removing cotter pin, remove the four screws attaching cover (35—Fig. 37) and solenoid to side of timing gear housing, then remove cover and solenoid. Remove pin (A—Fig. 38), then detach link (B) from pump pin. Remove the four nuts attaching pump to top of timing gear housing and withdraw pump.

Old shims (37—Fig. 37) should not be reinstalled, but new shims should be same thickness as originally installed. Thickness of shims should only be changed by specially trained personnel with proper special tools. Injection timing will occur later if thickness is increased. If original thickness is not known, manufacturer suggests that one new 0.5 mm (0.020 in.) thick shim be installed. Make sure that old paint and sealer are cleaned from mating surface of pump and timing gear housing.

Further disassembly of the pump is not recommended.

When installing, reverse removal procedure, using new shims (37) of same thickness as removed. If original thickness is not known, manufacturer suggests that one new 0.5 mm (0.020 in.) thick shim be installed. Tighten the four nuts retaining pump to 20 N·m (180 in.-lbs.) torque. Compression fittings of high pressure fuel lines should be tightened to 40 N·m (30 ft.-lbs.) torque. Bleed air from fuel system as outlined in paragraph 82.

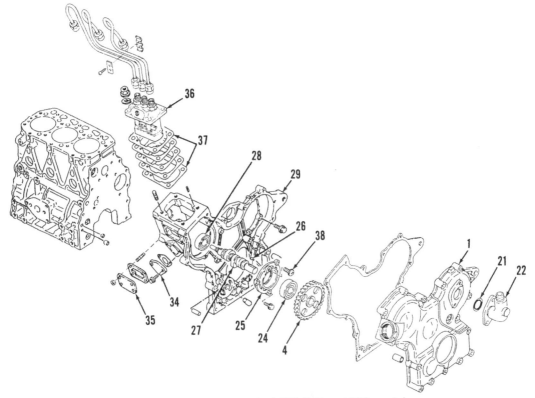

**Fig. 37—View of some injection pump components typical of 655, 755 and 756 models.**

| | | |
|---|---|---|
| 1. Timing gear cover | 24. Ball bearing | 35. Cover |
| 4. Injection pump drive gear | 25. Retainer | 36. Diesel injection pump |
| 21. "O" ring | 26. Drive key | 37. Shims |
| 22. Tachometer gear unit | 27. Injection pump camshaft | 38. Bearing retaining screw |
| | 28. Ball bearing | |
| | 29. Timing gear housing | |
| | 34. Cover | |

## 855, 856 And 955 Models

**86. TIMING TO ENGINE.** Timing of the injection pump camshaft and drive gear (4—Fig. 30) is considered correct if marks (T1 and T2) are aligned as shown. Specific injection timing can be changed by moving injection pump in the three elongated mounting slots, but timing should not be changed except by specially trained personnel. Original setting is as shown in Fig. 39. Tighten pump mounting nuts to 26 N·m (19 ft.-lbs.) torque.

**87. REMOVE AND REINSTALL.** To remove injection pump, clean pump and area around pump and lines. Remove banjo bolts (42—Fig. 40) and remove lubrication line (41). Turn fuel off at shut-off valve of

**Fig. 38—Pin (A) holds link (B) on pump pin.**

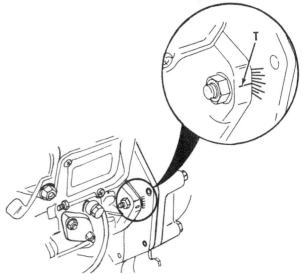

**Fig. 39—View of fuel injection pump timing marks for 855, 856 and 955 models.**

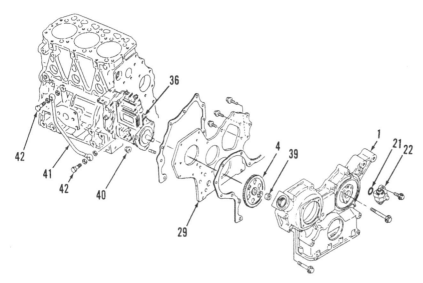

*Fig. 40—View of some injection pump components typical of 855, 856 and 955 models.*

1. Timing gear cover
4. Injection pump drive gear
21. "O" ring
22. Tachometer gear unit
29. Timing gear housing
36. Injection pump
39. Nut
40. Nut (3 used)
41. Oil line
42. Banjo bolts

filter housing, then disconnect fuel inlet line to pump and high pressure discharge lines to nozzles from all models. Cover all openings in pump, nozzles and fuel lines to prevent entrance of dirt. Remove cover from front of timing gear cover (1), then turn crankshaft until marked tooth valley of injection pump gear (4) is aligned as shown at (T2—Fig. 41).

**NOTE: Marked tooth of idler gear may not be aligned, because idler has uneven number of teeth.**

Mark tooth of idler gear, which is meshed with marked pump gear, with paint or chalk as shown to facilitate alignment, then remove nut (39).

**NOTE: Do not mark tooth of idler gear with punch or in another permanent method, because gear already has three timing marks (T1, T2 and T3—Fig. 30).**

Remove the three nuts (40—Fig. 40) attaching pump to timing gear housing, then use a suitable gear puller to push pump shaft from gear (4).

**NOTE: Do not crank engine with injection pump removed and gear loose in timing gear housing. If**

**temporary timing marks were not installed on idler gear as described, refer to paragraph 67 and remove timing gear cover. Refer to paragraph 62 and Fig. 30 for aligning timing marks on timing gears.**

When installing, reverse removal procedure, making sure that timing marks on gears remain aligned. Tighten the three nuts (40—Fig. 40) attaching pump to 27 N·m (20 ft.-lbs.) torque and gear retaining nut (39) to 88 N·m (65 ft.-lbs.) torque. Compression fittings of high pressure fuel lines should be tightened using two wrenches. Bleed air from fuel system as outlined in paragraph 82.

## GOVERNOR AND INJECTION PUMP CAMSHAFT

### 655, 755 And 756 Models

**88. REMOVE AND REINSTALL.** To remove the governor (fuel control) and linkage, clean pump and area around pump and governor. Detach solenoid linkage by removing cotter pin, remove the four screws attaching cover (35—Fig. 37) and solenoid to side of timing gear housing, then remove cover and solenoid. Remove pin (A—Fig. 38), then detach link (B) from pump pin. Unbolt and remove engine oil dip stick tube and disconnect speed control cable from lever (13—Fig. 42). Unbolt and remove bracket (1), remove four additional screws attaching governor housing (3) to rear of timing gear housing, then withdraw housing (3).

Inside diameter of sleeve (4) should not be worn larger than 8.20 mm (0.323 in.) and shaft OD should be no less than 7.90 mm (0.311 in.).

Remove nut (5), then remove flyweight assembly (6, 7 and 8). Further disassembly of governor assembly should only be accomplished by trained personnel. Minimum diameter of shaft (16) is 7.90 mm (0.311 in.)

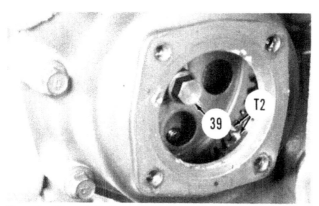

*Fig. 41—View of timing marks (T2) aligned on 855, 856 and 955 models. Because of uneven number of teeth on idler gear, mark on idler gear does not align with mark on pump gear every time.*

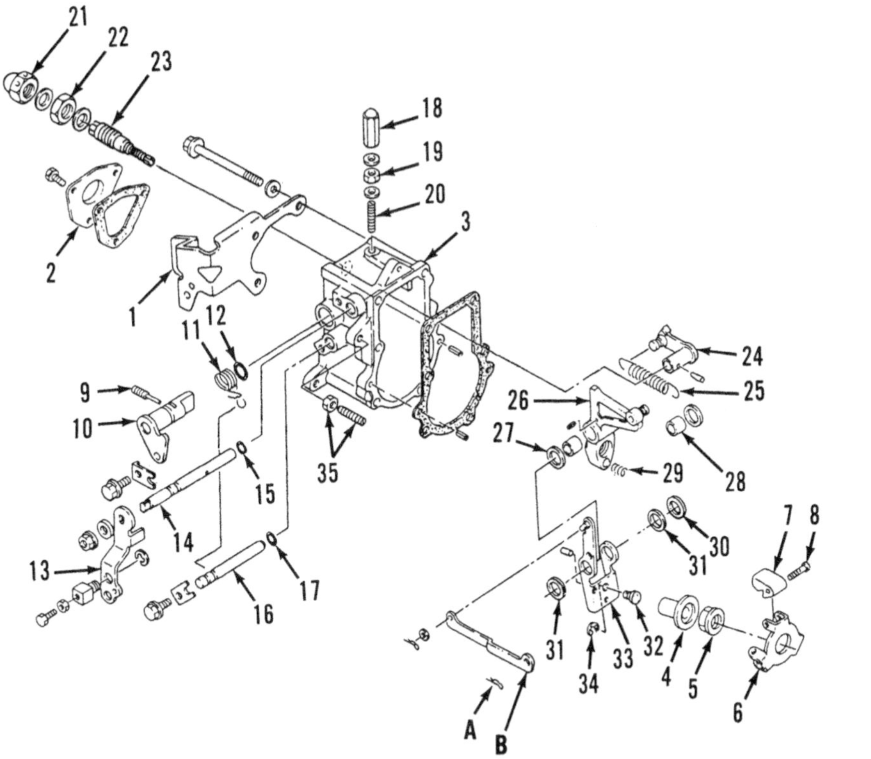

*Fig. 42—Exploded view of governor (fuel control) assembly typical of 655, 755 and 756 models.*

A. Pin
B. Link
1. Bracket
2. Cover
3. Housing
4. Sleeve
5. Nut
6. Weight retainer
7. Weight (3 used)
8. Pin
9. Retaining screw
10. Fuel shut-off lever
11. Return spring
12. "O" ring
13. Speed control lever
14. Shaft
15. "O" ring
16. Pivot shaft
17. "O" ring
18. Cap nut
19. Locknut
20. High speed
    stop screw
21. Cap nut
22. Locknut
23. Fuel control restrictor
24. Lever

25. Spring
26. Lever
27. Shim (2 used)
28. Bushing (2 used)

29. Spring
30. Shim
31. Washers
32. Shifter

33. Lever
34. "E" ring
35. Idle stop screw
    & locknut

and bushings (28) should not be worn larger than 8.15 mm (0.321 in.).

To remove the injection pump camshaft (27—Fig. 37), refer to preceding paragraphs and remove governor housing and weights. Refer to paragraph 36 and remove timing gear cover. Refer to paragraph 85 and remove injection pump. Remove bearing retaining screw (38), then pull camshaft (27), bearing (24) and drive gear (4) from timing gear housing (29) and bearing plate (25).

Install new injection pump camshaft if lobe height is less than 30.9 mm (1.217 in.).

When assembling, reverse removal procedure. Refer to Fig. 21 and paragraph 31 for aligning marks on all timing gears, including the injection pump drive gear. Tighten screw (38—Fig. 37) to 20 N·m (180 in.-lbs.) torque. Refer to paragraph 36 and install timing gear housing, paragraph 85 and install injection pump and to paragraph 89 for adjusting governor.

**89. ENGINE SPEED AND GOVERNOR ADJUSTMENTS.** Slow idle speed is adjusted by stop screw (35—Fig. 42) and high idle speed by stop screw (20). Settings of the high speed stop (20) and fuel restrictor adjustment (23) are sealed and should be changed only by trained personnel with proper equipment.

Before adjusting low or high idle engine speed stops, make sure that throttle control cable is correctly attached to lever (13) so that full movement is possible. Low idle speed should be 1400-1500 rpm and high idle no load speed should be 3400-3450 rpm.

## 855, 856 And 955 Models

**90.** The governor and injection pump camshaft are integral with the pump and disassembly is not recommended.

**91. ENGINE SPEED AND GOVERNOR ADJUSTMENTS.** Slow idle speed is adjusted by stop screw (35—Fig. 43) and high idle speed by stop screw (20). Setting of the high speed stop (20) is sealed and should be changed only by trained personnel with proper equipment.

Before adjusting low or high idle engine speed stops, make sure that throttle control cable is correctly attached to lever (13) so that full movement is possible. Low idle speed should be 1400-1500 rpm and high idle no load speed should be 3400-3450 rpm.

## INJECTOR NOZZLES

### All Models

**92. TESTING AND LOCATING A FAULTY NOZZLE.** If one or more cylinders is misfiring and

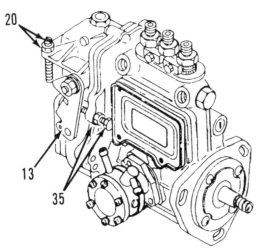

*Fig. 43—View of fuel injection pump, typical of 855, 856 and 955 models, showing governor (fuel control) adjustment points.*

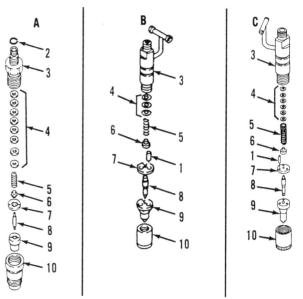

*Fig. 44—Type "A" injector nozzle is used in 655, 755 and 756 models; type "B" nozzle is used in 855 and 856 models; type "C" is used in 955 models.*

| | |
|---|---|
| 1. Alignment pins | 6. Spring seat |
| 2. "O" ring | 7. Stop plate |
| 3. Nozzle holder | 8. Nozzle needle |
| 4. Shims | 9. Nozzle body |
| 5. Spring | 10. Nozzle nut |

the fuel system is suspected as being the cause of trouble, check by loosening each injector line connection in turn, while the engine is running at slow idle speed. If the engine is not materially affected when an injector line is loosened, that cylinder is misfiring. Remove and test the nozzle or install a new or reconditioned unit as outlined in the appropriate following paragraphs.

**93. REMOVE AND REINSTALL.** Thoroughly clean injector, lines and surrounding area before loosening lines or removing injector. Detach compression fittings at both ends of high pressure lines, using two wrenches, then remove the lines. Disconnect or remove nozzle bleed (return) line from the nozzle or nozzles to be removed and cover all openings to prevent entrance of dirt.

On 655, 755 and 756 models, unscrew injector nozzle from cylinder head. Remove injector (A—Fig. 44), gaskets (11 and 13—Fig. 45) and seal (12) from injector bores in cylinder head.

On 855 and 856 models, remove nuts (25—Fig. 46), holder (26) and spacers (27), then pull injector (B—Fig. 44) from cylinder head. Remove seal (28—Fig. 46) and seat (29) from bore in cylinder head.

On 955 models, remove nuts (25—Fig. 47), holder (26) and spacers (27), then pull injector (C—Fig. 44) from cylinder head. If nozzle is difficult to remove from cylinder head, it may be necessary to use a puller attached to the threads for the high pressure fuel line. **Be careful not to damage nozzle while attempting to remove.** Remove seal (28—Fig. 47) from bore in cylinder head.

Before installing injectors on all models, first be sure that bore in cylinder head is clean and all seals

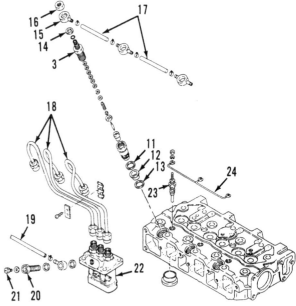

*Fig. 45—View of injection pump and nozzle for 655, 755 and 756 models. Nozzle is shown exploded at "A" in Fig. 44.*

| | |
|---|---|
| 3. Nozzle holder | |
| 11. Gasket | 18. High pressure lines |
| 12. Seal | 19. Supply line |
| 13. Gasket | 20. Fitting |
| 14. Gasket | 21. Bleeder fitting |
| 15. Banjo fitting | 22. Injection pump |
| 16. Nut | 23. Glow plug |
| 17. Return lines | 24. Connector |

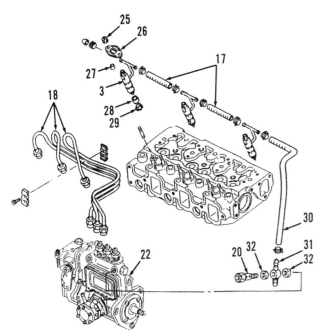

**Fig. 46—View of injection pump and nozzle for 855 and 856 models. Nozzle is shown exploded at "B" in Fig. 44.**

| | |
|---|---|
| 3. Nozzle holder | 27. Spacer |
| 17. Return lines | 28. Seal |
| 18. High pressure lines | 29. Seat |
| 20. Fitting | 30. Return line |
| 22. Injection pump | 31. Banjo fitting |
| 25. Nut | 32. Gaskets |
| 26. Clamp | |

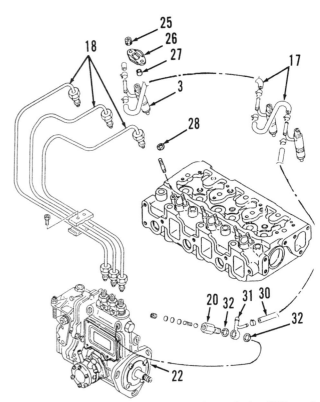

**Fig. 47—View of injection pump and nozzle for 955 models. Nozzle is shown exploded at "C" in Fig. 44.**

| | |
|---|---|
| 3. Nozzle holder | 26. Clamp |
| 17. Return lines | 27. Spacer |
| 18. High pressure lines | 28. Seal |
| 20. Fitting | 30. Return line |
| 22. Injection pump | 31. Banjo fitting |
| 25. Nut | 32. Gaskets |

are removed. Clean the bore in cylinder head carefully and completely, then inspect bore for both cleanliness and damage. Insert new gaskets and seals, then install injector in bore.

On 655, 656, 755 and 756 models, tighten injector to 50 N·m (37 ft.-lbs.) torque in cylinder head and tighten nut (16—Fig. 45) to 40 N·m (30 ft.-lbs.) torque.

On 855 and 856 models, tighten injector hold down nuts (25—Fig. 46) to 4.5 N·m (39 in.-lbs.) torque in cylinder head.

On 955 models, tighten injector hold down nuts (25—Fig. 47) to 4.5 N·m (39 in.-lbs.) torque in cylinder head.

On all models, refer to paragraph 82 and bleed the fuel system. Bleed high pressure lines (18—Fig. 45, Fig. 46 or Fig. 47) before tightening line fittings to injectors.

**94. TESTING.** A complete job of testing, cleaning and adjusting the fuel injector requires removal as outlined in paragraph 93 and the use of special test equipment. Use only clean, approved testing oil in the tester tank. Injector should be tested for opening pressure, seat leakage and spray pattern. Before connecting the injector to the test stand, operate

tester lever until oil flows, then attach injector to tester line.

> **WARNING: Fuel leaves injector nozzle with sufficient force to penetrate the skin. Keep exposed portions of your body clear of nozzle spray when testing.**

OPENING PRESSURE. Open valve to tester gauge and operate tester lever slowly while observing gauge reading. Opening pressure should be 11,242-12,202 kPa (1630-1770 psi) for 655, 755 and 756 models. Opening pressure should be 19,120-20,080 kPa (2773-2913 psi) for 855, 856 and 955 models. Opening pressure is changed by adding or removing shims (4—Fig. 44). Set opening pressure to upper limit if opening pressure is changed.

SPRAY PATTERN. Operate the tester lever slowly and notice the spray from the nozzle. Nozzle should emit a chattering sound and should spray a fine stream. Operate tester handle fast and again observe the spray from nozzle. Spray should be finely atomized and the pattern should be even. If nozzle does not chatter, spray is not finely atomized or sprays to one side, disassemble and clean or renew the nozzle.

LEAKAGE TEST. Wipe end of nozzle with a clean cloth after completing opening and spray pattern tests.

On 655, 755 and 756 models, operate tester until pressure reaches pressure of 11,032 kPa (1600 psi) and maintain this pressure. Nozzle should not drip for at least 10 seconds; however, end of nozzle may become wet.

On 855, 856 and 955 models, operate tester until pressure reaches pressure of 17,640 kPa (2550 psi) and maintain this pressure. Nozzle should not drip for at least 5 seconds; however, end of nozzle may become wet.

On all models, leakage can usually be corrected by cleaning.

**95. OVERHAUL.** Do not use hard or sharp tools, emery cloth, grinding compound or other than approved solvents or lapping compounds. Approved nozzle cleaning kits are available from specialized tool sources.

Wipe all dirt and loose carbon from exterior of nozzle and holder assembly. Refer to Fig. 44 for exploded view and proceed as follows:

Secure nozzle in a soft-jawed vise or holding fixture and remove nut (3). Place all parts in clean calibrating oil or diesel fuel as they are removed. Use a

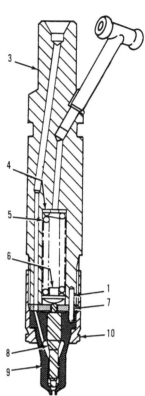

Fig. 49—Cross section of injector used on 855, 856 and 955 models. Refer to Fig. 44 for legend.

compartmented pan and exercise care to keep parts from each injector together and separate from other units that are disassembled at the same time.

Clean exterior surfaces with a brass wire brush, soaking parts in an approved carbon solvent if necessary, to loosen hard carbon deposits. Rinse parts in clean diesel fuel or calibrating oil immediately after cleaning to neutralize the solvent and prevent etching of polished surfaces.

Clean nozzle spray hole from inside using a pointed hardwood stick or wood splinter, then scrape carbon from pressure chamber using a hooked scraper. Clean valve seat using brass scraper, then polish seat using wood polishing stick and mutton tallow. Refer to Fig. 48 and Fig. 49 for cross-sectional views of injectors.

Reclean all parts, rinsing thoroughly in clean diesel fuel or calibrating oil and assemble while parts are immersed in clean fluid. Make sure adjusting shim pack is intact. Tighten nozzle retaining nut to 40 N·m (30 ft.-lbs.) torque for 655, 755 and 756 models. Tighten nozzle retaining nut to 43 N·m (31 ft.-lbs.) torque for 855, 856 and 955 models. Do not overtighten. Distortion may cause valve to stick and no amount of overtightening can stop a leak caused by scratches or dirt. Retest assembled injector as previously outlined.

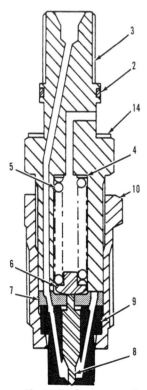

Fig. 48—Cross section of injector used on 655, 755 and 756 models. Refer to Fig. 45 and Fig. 44 for legend.

## PREHEATER / GLOW PLUGS

### All Models

**96.** On 655, 755 and 756 models, three glow plugs (23—Fig. 45) are connected in parallel with each glow plug grounding through mounting threads. A preheater is located at the inlet of engine inlet manifold of 855, 856 and 955 models. On all models, starting (key) switch is provided with "preheating" position which can be used to energize the glow plugs or preheater for faster warm up. When starting (key) switch of 655, 755 and 756 models is turned to the first (ON) position, the glow plugs and preheat indicator light on dash will be energized for approximately 8 seconds. On 855, 856 and 955 models, the first (ON) position of the starting (key) switch will energize preheater and indicator light for approximately 18 seconds if the ambient temperature is below 14° C (58° F).

If preheater indicator lamp fails to light when start switch is first in "ON" position and at appropriate ambient temperature, check for loose connections at switch, indicator lamp, resistor connections, glow plug (or preheater) connections and ground. A test lamp can be used to check for current flow to glow plugs or preheater. Red wire to preheater control module is equipped with a fusible link. The temperature sensitive preheater control module for 855, 856 and 955 models can be checked for operation when weather is warm by removing and cooling the control module in a freezer, then quickly installing and testing.

# COOLING SYSTEM

## RADIATOR

### All Models

**97.** Radiator pressure cap is used to increase cooling system pressure. Valve in cap is set to open at about 97-104 kPa (14-15 psi). Cooling system capacity is 3.8 L (4 qt.) for 655, 755, 756, 855 and 856 models; 4.5 L (4.8 qt.) for 955 models.

To remove radiator from all models, drain coolant, remove grille, engine side panels, hood, and air cleaner inlet hose. Detach upper and lower hoses from radiator.

The fuel tank is located in front of radiator on all 655, early 755, all 756, early 855 and all 856 models. Disconnect the two fuel lines that are located at lower right side of radiator and route the hoses forward, out of radiator support. Disconnect fuel level wires from the sender in tank and route the wires toward rear, out of radiator support.

The battery is located in front of the radiator of late 755, late 855 and all 955 models. Disconnect cables from battery, remove battery, then route cables toward rear, out of radiator support.

On all models, unbolt radiator/hood support from the top front radiator brackets. Remove screws from lower end of radiator/hood support. Pry the fuel tank, if so equipped, forward, then remove radiator forward and up, away from fan.

Reassemble by reversing removal procedure. Clearance between fan blades and radiator must be at least 12 mm (½ in.) for all models. Center to center distance between side panel mounting tabs should be 508 mm (20 in.) for 655 models; 562 mm (22.1 in.) for 755 and 756 models. Center to center distance between side panel mounting tabs should be 567 mm (22.3 in.) for 855, 856 and 955 models.

## THERMOSTAT

### All Models

**98.** The thermostat (18—Fig. 50, Fig. 51, Fig. 52 or Fig. 53) is located in the coolant outlet and should begin to open at 71° C (160° F). Install new thermostat if not completely open at 85° C (184° F).

Tighten screws attaching coolant outlet to 9 N·m (78 in.-lbs.) torque on 655 models, 26 N·m (226 in.-lbs.) torque for 755 and 756 models, or 20.3 N·m (180 in.-lbs.) torque for 855, 856 and 955 models.

## WATER PUMP

### All Models

**99.** To remove water pump, first drain coolant and remove the radiator. Loosen alternator attaching bolts, then remove fan, water pump and alternator drive belt. Unbolt and remove fan (1—Fig. 50, Fig. 51, Fig. 52 or Fig. 53). Remove pump mounting screws and separate pump from engine.

Pump repair parts are available from manufacturer, but installation of new pump is often more satisfactory. Measure distance from front face of hub (4) to rear of pump shaft (5) before disassembling so that hub can be installed to correct dimension. Press only on outer edge of shaft bearing (5) and press into housing until front of bearing is flush with front of housing. Press seal (8) into housing until it is bottomed in bore, using a tool that contacts only the outer edge of seal.

Press hub (4) onto pump shaft (5) until flush with end of shaft for 655, 755 and 756 models or until

bottom of flange is 17 mm (0.67 in.) from top of housing for 855, 856 and 955 models. On all models, install ceramic part of seal (8) in impeller, then press

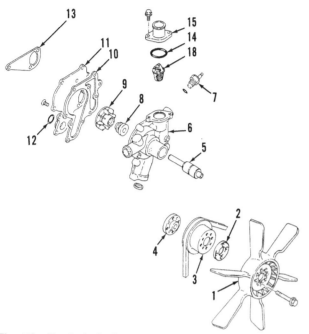

*Fig. 50—Exploded view of water pump typical of 655 models.*

1. Fan
2. Spacer
3. Pulley
4. Hub
5. Pump shaft & bearing
6. Pump housing
7. Temperature sensor
8. Seal
9. Impeller
10. Gasket
11. Cover
12. Seal
13. Gasket
14. Seal
15. Outlet
18. Thermostat

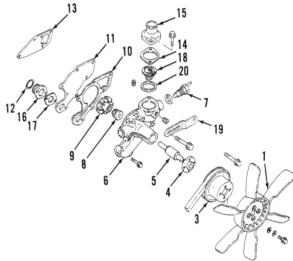

*Fig. 52—Exploded view of water pump typical of 855 and 856 models.*

1. Fan
3. Pulley
4. Hub
5. Pump shaft & bearing
6. Pump housing
7. Temperature sensor
8. Seal
9. Impeller
10. Gasket
11. Cover
12. Seal
13. Gasket
14. Seal
15. Outlet
16. Pipe
17. Gasket
18. Thermostat
19. Alternator bracket
20. Gasket

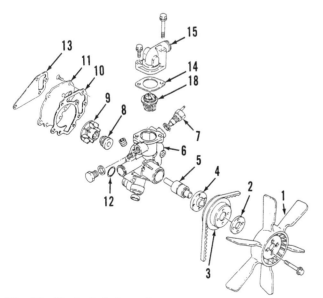

*Fig. 51—Exploded view of water pump typical of 755 and 756 models. Refer to Fig. 50 for legend.*

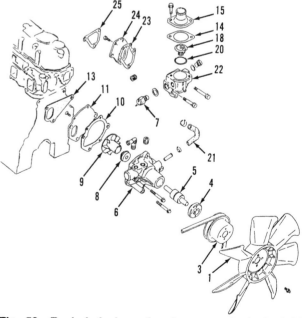

*Fig. 53—Exploded view of water pump typical of 955 models. Refer to Fig. 52 for legend except the following.*

21. Hose
22. Thermostat housing
23. Gasket
24. Plate
25. Gasket

impeller (9) onto shaft until front surface of impeller is 0.3-1.1 mm (0.012-0.043 in.) from pump housing (6). Rear of impeller will be approximately 1-2 mm (0.04-0.08 in.) below flush with rear of housing. Install gasket (10) and rear plate (11), tightening retaining screws to 9 N·m (78 in.-lbs.) torque. Screws retaining fan (1) and pulley (3) to front hub should be tightened to 11 N·m (96 in.-lbs.) torque. Screws attaching pump to engine should be tightened to 26

N·m (226 in.-lbs.) torque. Install pump in reverse of removal procedure.

Location, style and thickness of spacer (2) and hub (4) locates fan and pulley. Be sure pulley (3) is aligned with alternator and crankshaft pulleys. Tighten drive belt until approximately 13 mm (½ in.) slack can be measured at midpoint between alternator pulley and crankshaft pulley, when pressing with moderate thumb pressure.

# ELECTRICAL SYSTEM

## ALTERNATOR AND REGULATOR

**100.** Nippon Denso 35 amp alternator is used on early models and Nippon Denso 40 amp alternator is used on later models. Both models include an integral voltage regulator .

## Nippon Denso 35 Amp

**101.** To service the voltage regulator and brushes, remove the three through-bolts from front of alternator and the nuts and insulators from rear of alterna-

tor. Separate between stator (12—Fig. 54) and rear housing (14).

**NOTE: Do not heat soldered connections longer than necessary. Soldering gun must have at least 120 watt capacity.**

Unsolder leads, then remove brushes (13) and voltage regulator (16) as necessary. Make sure that brush springs are in good condition. Install new brushes if exposed length of brushes is less than 5.5 mm (0.220 in.). Exposed length of new brushes is 13 mm (1/2 in.). Brushes must move freely in holder.

Use only 60-40 rosin core solder when reconnecting. Brushes can be held up with wire inserted through hole in back while installing rear housing. Be sure to remove wire after assembling.

To disassemble alternator, first remove nut (20—Fig. 54), then use suitable puller to remove pulley (1). Remove fan (2), then press or bump shaft of rotor (10) from bearing (7). Bearing can be removed from front housing after removing retainer plate (8).

Stator (12—Fig. 54) must be unsoldered to test. Low resistance should be noticed between each pair of three stator wires, but wires should not be grounded to stator frame (core). Resistance between each pair of stator leads should be the same as the other two pairs. Check stator for discoloration caused by overheating and other visual signs of damage.

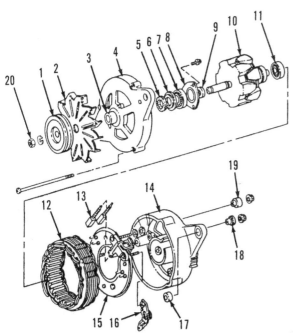

**Fig. 54—Exploded view of Nippon Denso 35 amp alternator used on early models.**

| | |
|---|---|
| 1. Pulley | 11. Bearing |
| 2. Fan | 12. Stator |
| 3. Locking collar | 13. Brushes |
| 4. Drive end frame | 14. Stator housing |
| 5. Felt washer | 15. Rectifier (diode) assy. |
| 6. Cover | 16. Voltage regulator |
| 7. Bearing | 17. Bushing |
| 8. Retainer plate | 18. Insulator |
| 9. Locking collar | 19. Output terminal bushing |
| 10. Rotor | 20. Nut |

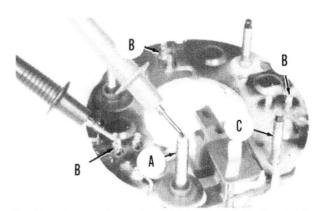

**Fig. 55—Use an ohmmeter to test rectifier diodes. Refer to text for procedure.**

Rectifier (diode) assembly can be checked with ohmmeter after separating from stator. First, test continuity between each of the three outside terminals (B—Fig. 55) and output post (A) of alternator. Reverse polarity of test by reversing ohmmeter leads. Continuity should exist with one connection, but not when test leads are reversed. Second, test continuity between each of the outside terminals (B) and the ground post (C). Reverse polarity of test by reversing ohmmeter leads. Continuity should exist with one connection, but not when test leads are reversed. Install new rectifier if any test indicates a diode is shorted, by allowing current to pass with either ohmmeter connection or if any test indicates a diode is open, by not allowing current to pass either direction.

Reverse disassembly procedure when reassembling. Use only 60-40 rosin core solder when resoldering and soldering gun should have at least 120 watt capacity. Do not heat any connection longer than necessary, especially near the rectifier which is easily damaged by heat. Be sure that stator lead wires do not contact housing frame when assembled. Hold brushes up with wire inserted through hole in back while installing rear housing and be sure to remove

wire after assembling. Tighten pulley retaining nut (20—Fig. 54) to 54 N·m (40 ft.-lbs.) torque.

## Nippon Denso 40 Amp

**102.** To disassemble alternator, refer to Fig. 56 and remove end cover (15). Remove through-bolts and separate drive end frame (2) from stator housing (10). Remove nut from rotor shaft and tap rotor (6) from end frame. Remove retainer plate (5) and bearing (4). Unsolder voltage regulator and brush holder wire leads and separate voltage regulator (12) from brush holder (13). Unsolder stator wire leads and remove rectifier assembly (11).

Use an ohmmeter to check rotor winding for short circuit or open circuit. There should be continuity when tester leads are connected to rotor slip rings. There should not be continuity when one of the tester leads is connected to rotor frame and the other is attached to slip rings. Install new brushes (14) if exposed length is less than 4.5 mm (0.18 in.). Exposed length of new brush is 10.5 mm (0.41 in.). Check for continuity between each of the brushes and between each brush and ground. Renew brush holder (13) if continuity is indicated.

Inspect stator (3) for discoloration or burned odor, which would indicate faulty wiring, and renew as necessary. There should be equal resistance readings between each pair of stator lead wires. There should not be continuity indicated between any of stator leads and stator frame.

To test rectifier diode, connect tester leads to output terminal (A—Fig. 57) and to each of the diode leads (B), then reverse tester leads and repeat test. There should be continuity indicated in only one direction. Renew rectifier assembly if any of the diodes fails the test.

To reassemble alternator, reverse the disassembly procedure. When resoldering lead wires, use only 60-40 rosin core solder. Do not heat any connection longer than necessary, especially near the rectifier which is easily damaged by heat. Be sure that stator

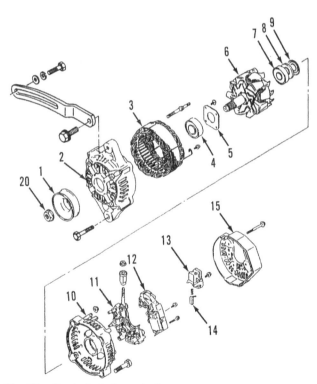

*Fig. 56—Exploded view of 40 amp alternator used on late models.*

1. Pulley
2. End frame
3. Stator
4. Bearing
5. Retainer plate
6. Rotor
7. Bearing
8. Thrust washer
9. Cover
10. Housing
11. Rectifier assy.
12. Voltage regulator
13. Brush holder
14. Brush
15. End cover

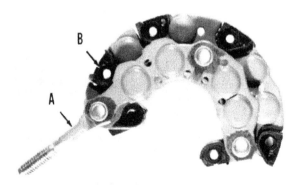

*Fig. 57—Use an ohmmeter to test rectifier diodes. Refer to text for procedure.*

lead wires do not contact housing frame when assembled. Hold brushes up with wire inserted through hole in back while installing rear housing and be sure to remove wire after assembling. Tighten pulley retaining nut (20—Fig. 56) to 69 N·m (51 ft.-lbs.) torque.

## STARTER

### All Models

**103.** Model 655 tractors are equipped with Hitachi 0.8 kW starter and other models are equipped with 1.0 kW starter manufactured by Nippon Denso. Refer to the appropriate following paragraphs for tests and service.

### Hitachi (0.8 kW)

**104.** To disassemble, disconnect solenoid wire, then remove through-bolts and retaining screws from end cover. Pry plastic cap (18—Fig. 58) from end cover, remove "E" ring (17), shims (16) and end cover (14).

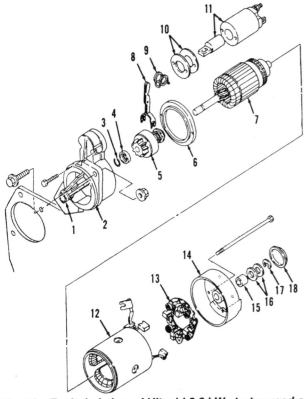

*Fig. 58—Exploded view of Hitachi 0.8 kW starter used on 655 models.*

| | |
|---|---|
| 1. Sleeve | 10. Shims |
| 2. Drive end frame | 11. Solenoid assy. |
| 3. Snap ring | 12. Field coil |
| 4. Pinion stop | 13. Brush holder |
| 5. Drive pinion & clutch | 14. End cover |
| 6. Cover | 15. Sleeve |
| 7. Armature | 16. Shims |
| 8. Shift lever | 17. "E" ring |
| 9. Spring | 18. Cap |

Pry springs away from ground brushes, pull brushes up, then release springs so that springs will hold brushes up. Remove field coil brushes from brush holder, then remove brush holder (13) from field housing. Remove the field coil housing (12). Remove the two screws attaching starter solenoid (11) to drive end frame (2), then separate solenoid and armature from drive end frame. Push pinion stop (4) downward, remove retaining ring (3) and withdraw clutch assembly (5) from armature shaft.

Inspect all components for wear or damage. Install new brushes if less than 7.7 mm (0.30 in.) long. Use an ohmmeter to test armature for grounded or open windings. Check for shorts between commutator segments and armature shaft. Continuity should exist between two commutator segments. An armature growler should be used to check condition of armature. Field winding can be checked with an ohmmeter. Check for short circuit between field coil brush and field frame. Continuity should exist between the two field brushes.

Reassemble starter by reversing disassembly procedure. Be sure to seat pinion stop (4) over retaining ring (3) when assembling. Rubber plug is used to seal between solenoid and drive end frame. Plastic cap (18) seals and protects end of armature shaft, "E" ring and washers.

### Nippon Denso (1.0 kW)

**105.** To disassemble, disconnect field lead, remove through-bolts and separate starter motor from solenoid housing. Remove end frame (1—Fig. 59). Pull brush springs away from brushes and lift field coil brushes from brush holder. Pry springs away from ground brushes, pull brushes up about 6 mm (¼ in.), then release springs so that springs will hold brushes up. Separate brush holder (2) from field coil housing (3). Withdraw armature (5) from field housing.

To disassemble reduction gears and clutch assembly, remove retaining screws and separate solenoid housing (10) from clutch housing (25). Remove clutch assembly and drive pinion from clutch housing. Push retainer (23) back and remove retainer ring (24), then withdraw components from clutch shaft (16).

Remove cover (7) from solenoid housing and withdraw plunger (8). Remove contact plates (9) if necessary. Notice that contact plate located on left side of housing is smaller than contact plate on right side.

Inspect all components for wear or damage. Install new brushes if length is less than 8.5 mm (0.30 in.).

Use an ohmmeter to test for grounded or open winding. Check for shorts between commutator segments and armature shaft. Continuity should exist between two commutator segments. An armature growler should be used to check condition of armature. Field winding can be checked with an ohmmeter. Check for short circuit between field coil brush

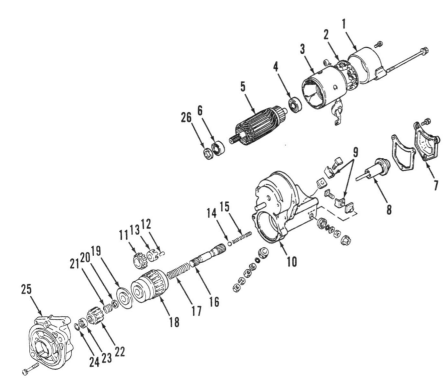

**Fig. 59—Exploded view of Nippon Denso starter typical of type used on all models except 655.**

1. End frame
2. Brush holder
3. Field coil frame
4. Bearing
5. Armature
6. Bearing
7. Cover
8. Plunger
9. Contact plates
10. Solenoid housing
11. Pinion gear
12. Roller
13. Retainer
14. Ball
15. Spring
16. Clutch shaft
17. Spring
18. Clutch assy.
19. Washer
20. Toothed retainer washer
21. Spring
22. Drive gear
23. Retainer
24. Snap ring
25. Clutch housing
26. Felt washer

*Fig. 60—Drawing of hydrostatic controls. Neutral start switch is attached to hydro swashplate bracket.*

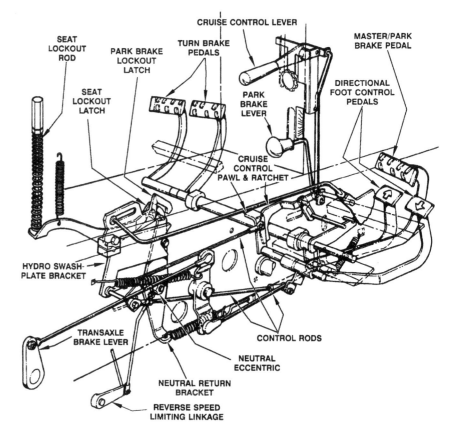

and field frame. Continuity should exist between the two field brushes.

Reassemble starter by reversing disassembly procedure while observing the following. Grease bearings of clutch assembly (18—Fig. 59). Install spring (17), clutch assembly (18) and washer (19) onto clutch shaft (16), then install retainer (20) and turn retainer to hold parts on shaft. Install spring (21) and drive gear (22), compress parts (16 through 22) in a vise, then install retainer (23) and snap ring (24). Slide retainer (23) over snap ring and release parts from vise. Apply grease to rollers (12), retainer (13) and pinion gear (11), then install parts (11, 12 and 13) on shaft of clutch housing (25) at the same time clutch and shaft assembly (16 through 24) is installed in clutch housing (25). Install steel ball (14) into clutch shaft (16), using grease to hold bearing in place. Install spring (15) in bore of solenoid housing (10) and assemble solenoid housing (10) and clutch housing (25). Attach housings (10 and 25) together with the two screws. Grease felt washer (26) before assembling. Be sure that field coil brush wires do not contact end frame when installing end frame (1). Tighten the two through-bolts to 88 N•m (65 ft.-lbs.) torque.

## SAFETY SWITCHES

### All Models

**106.** A safety switch is provided to prevent starting unless the transmission is in neutral. Another safety switch is provided which prevents operation of the pto unless an operator is located on the operator seat. Refer to the appropriate wiring diagram for schematic of the safety circuits. The neutral start switch is attached to the hydro swashplate bracket (Fig. 60). Be sure to reinstall spacer between switch and bracket.

## CIRCUIT DESCRIPTION

### All Models

**107.** Refer to Fig. 61, Fig. 62, Fig. 63 or Fig. 64 for appropriate wiring diagram. The two fusible links in red wires near starting motor solenoid are 1.0 mm diameter. The light switch positions are: No terminals are connected when in "OFF" position; "WARN"

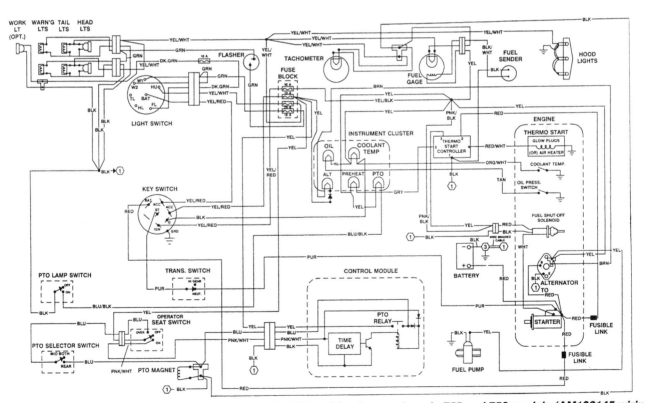

*Fig. 61—Wiring schematic for early 655 models (AM102054 wiring harness), early 755 and 756 models (AM102145 wiring harness), early 855 and 856 models (AM102053 wiring harness). Terminals (1) are grounded to engine and terminal (3) is grounded to tractor chassis.*

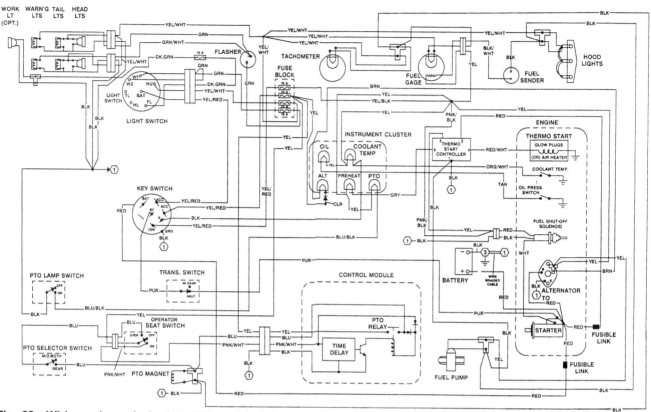

**Fig. 62—Wiring schematic for late 655 models (AM105090 wiring harness), intermediate 755 models (AM10591 wiring harness), and intermediate 855 models (AM10592 wiring harness). Terminals (1) are grounded to engine and terminal (3) is grounded to tractor chassis.**

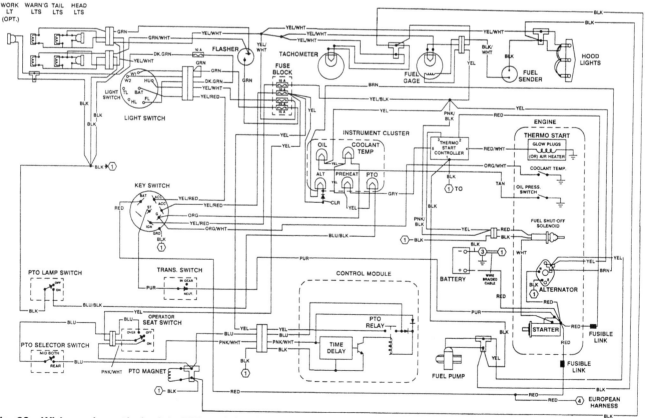

**Fig. 63—Wiring schematic for late 755 models with AM108177 wiring harness. Terminals (1) are grounded to engine and terminal (3) is grounded to tractor chassis. Terminal (4) is not used on most models.**

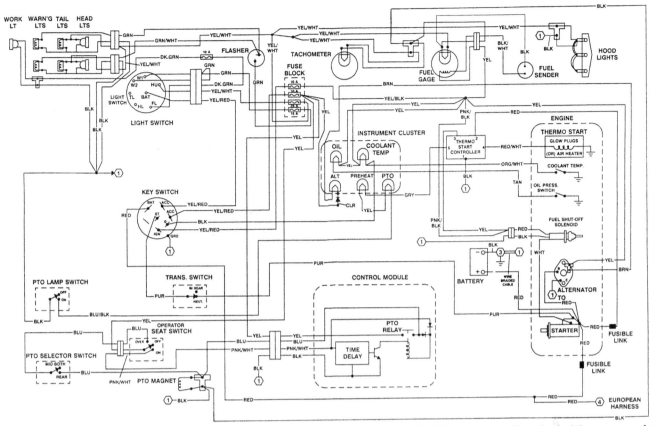

*Fig. 64—Wiring schematic for late 855 models and 955 models with AM106740 wiring harness. Terminals (1) are grounded to engine and terminal (3) is grounded to tractor chassis. Terminal (4) is not used on most models.*

which connects "BAT" and "W1" terminals; "WORK" which connects "BAT, HU" and "FL" terminals. The key switch positions are: "ACC" which connects "BAT" and "ACC" terminals; no terminals are connected when in "OFF" position; "RUN" which connects "BAT, IGN" and "ACC" terminals; "START" which connects "BAT, IGN" and "ST." Terminals "G" and "GND" are also bridged when key switch is in "START" position, which causes coolant temperature warning light to glow when key is in starting position.

# TRANSMISSION

**108.** The hydrostatic transmission consists of a variable displacement piston type hydraulic pump and a fixed displacement piston type hydraulic motor. Refer to Fig. 65 for schematic of hydrostatic fluid circuit. The hydrostatic pump is driven at engine speed by the input shaft, which is driven directly by the drive shaft from engine. The hydrostatic unit has two output shafts. The upper shaft provides drive from engine to pto upper shaft. The lower shaft is the hydrostatic motor output shaft, which drives the "Hi/Lo" range transmission gears located in rear axle center housing.

The hydrostatic drive provides variable speeds, both forward and reverse, in two speed ranges. Ground speeds should be within the following ranges:

**655 Models**
Forward speed
Low range . . . . . . . . . . . 0-8.7 km/h (0-5.4 mph)
High range . . . . . . . . 0-16.1 km/h (0-10.0 mph)
Reverse speed
Low range . . . . . . . . . . 0-4.3 km/h (0-2.7 mph)
High range . . . . . . . . . 0-8.0 km/h (0-5.0 mph).

**755 Models**
Forward speed
Low range . . . . . . . . . . 0-9.3 km/h (0-5.8 mph)
High range . . . . . . . . 0-17.1 km/h (0-10.6 mph)
Reverse speed
Low range . . . . . . . . . . 0-4.7 km/h (0-2.9 mph)
High range . . . . . . . . . 0-8.5 km/h (0-5.3 mph).

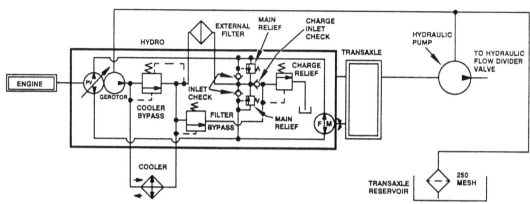

*Fig. 65—Schematic of hydrostatic transmission system used on all models. Fluid from main hydraulic pump is used for steering, pto, lift arms and remote hydraulics.*

## 855 Models

Forward speed
Low range . . . . . . . . . . . . 0-9.7 km/h (0-6.0 mph)
High range . . . . . . . . 0-17.7 km/h (0-11.0 mph)
Reverse speed
Low range . . . . . . . . . . 0-4.8 km/h (0-3.0 mph)
High range . . . . . . . . . 0-8.9 km/h (0-5.5 mph).

## 955 Models

Forward speed
Low range . . . . . . . . . . 0-8.3 km/h (0-5.1 mph)
High range . . . . . . . . 0-18.3 km/h (0-11.4 mph)
Reverse speed
Low range . . . . . . . . . 0-8.2 km/h (0-5.1 mph)
High range . . . . . . . . 0-9.2 km/h (0-5.7 mph).

## LUBRICATION

## All Models

**109.** The hydrostatic unit and hydraulic system utilize the same fluid contained in rear axle center housing. Fluid and canister filter for the hydraulic system and hydrostatic transmission should be changed and the transmission vent tube (1—Fig. 66) and screen (8) should be removed and cleaned after 500 hours of operation.

Lower the three-point hitch and remove drain plugs (3) from sides and rear (4) of housing. The rear drain plug (4) is located at rear of adapter (5), which is attached to rear of axle center housing, and the drawbar should be removed before removing this rear plug. After oil has drained, detach hose (10), remove the three screws attaching adapter (5), then remove adapter. Pull screen (8) from housing bore, clean in a suitable solvent, dry screen thoroughly, then reinstall. Remove and discard canister filter (Fig. 67), then install new filter (John Deere part number

AM102723 or equivalent). Oil should be changed, canister filter renewed and filter screen cleaned at the same time. Clean canister filter adapter and lubricate filter seal with clean oil before installing new filter. Tighten filter by hand, do not use filter removal wrench to install. Unscrew "U" shaped vent (1—Fig. 66) from top of rockshaft housing (12), located next to fill plug (2), clean vent with suitable solvent, then reinstall vent. Fill reservoir with "John Deere Low Viscosity HY-GARD" transmission and hydraulic oil or equivalent, then start engine. Allow engine to run at idle while operating hydraulic controls and raising and lowering lift arms. Stop engine, lower three-point hitch and check fluid level. Tractor should be on level surface when checking fluid level. Dipstick (9) is located in pipe threaded into rear cover (17) of rear axle center housing. Plug (2) for filling axle center housing and hydraulic system is located in top rear of rockshaft housing. Capacity is approximately 17 L (18 quarts), but some oil may remain in system when draining. Control linkage should be lubricated with lithium base multipurpose grease.

It may be necessary to bleed air from transmission, hydraulic and steering systems as follows after draining the system or installing a dry repair part. Fill the system, start engine and operate at high idle speed for one minute. Turn steering wheel full left and hold for five seconds, turn steering wheel until wheels point straight forward and hold for ten seconds, then turn steering wheel to full right position and hold for five seconds. Return wheels to straight ahead position and move tractor forward about 6 meters (20 ft.), make two hard left turns, then make two hard right turns. Move tractor about 3 meters (10 ft.) in reverse, then shut off engine and inspect for oil leaks. Recheck oil level and fill to correct level "John Deere Low Viscosity HY-GARD" transmission and hydraulic oil or equivalent.

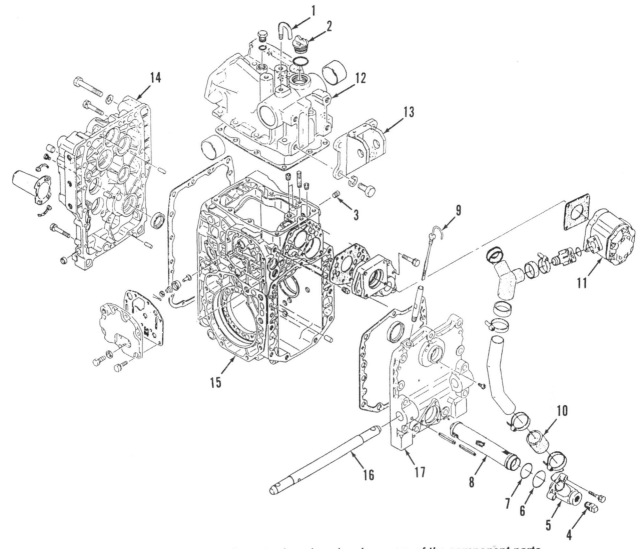

**Fig. 66—Partially exploded view of transaxle center housing showing some of the component parts.**

1. Vent
2. Fill plug
3. Side drain plug
4. Rear drain plug
5. Adapter
6. Back-up ring
7. "O" ring
8. Filter screen
9. Dipstick
10. Hose
11. Hydraulic pump
12. Rockshaft housing
13. Upper link bracket
14. Transaxle housing front cover
15. Transaxle housing
16. Lower link pin
17. Transaxle housing rear cover

## TROUBLESHOOTING

### All Models

**110.** The following are symptoms which may occur during operation and possible causes related to the hydrostatic transmission.

1. Tractor will not move or operates erratically.
   a. Transmission oil level low.
   b. Control linkage improperly adjusted or damaged.
   c. Charge pump damaged.
   d. Input drive shaft or coupling failure.

**Fig. 67—Canister filter shown is located on lower front surface of hydrostatic unit. Use filter wrench only for removal, not to install new filter.**

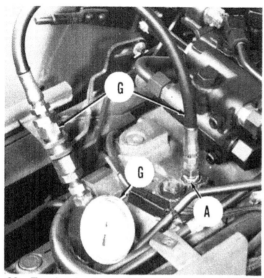

*Fig. 68—To check charge pump pressure, remove plug from port at (A) and install gauge (G) as shown.*

2. Tractor moves, but lacks power.
   a. Suction screen plugged.
   b. Charge pressure too low.
   c. Main relief valve damaged or spring broken.
   d. Internal leakage of high pressure oil.

3. Abnormal noise when operating.
   a. Suction filter plugged.
   b. Air leak on suction side of charge pump.
   c. Charge pump or relief valve faulty.

4. Tractor fails to stop in neutral.
   a. Control linkage out of adjustment.

5. Oil overheating.
   a. Tractor overloaded.
   b. Oil cooler plugged.

6. Tractor moves in only one direction.
   a. Control linkage worn, damaged or not adjusted properly.
   b. Main relief check valve or spring damaged.
   c. Internal leakage or damage.

## TESTS AND ADJUSTMENTS

### All Models

**111. CHARGE PRESSURE.** Remove platform from transmission housing and clean area thoroughly. Apply parking brake, block wheels and disengage front-wheel drive (if so equipped). Make sure fluid is at proper level and install test gauge as shown in Fig. 68. Start and run engine until fluid reaches normal (43° C – 110° F) operating temperature, then set engine to run at fast idle speed and observe gauge pressure. Correct charge pressure is 965-1379 kPa (140-200 psi). If charge pressure is incorrect, remove relief valve (43—Fig. 69) and inspect for damage. Pressure can be adjusted by adding or removing shims (44) located between spring and plug. Each 0.03 mm (0.001 in.) shim will change pressure about 7 kPa (1 psi).

**112. CHARGE PUMP FLOW.** Flow of charge pump can be checked by attaching a flow meter to the connections for the oil cooler at front of tractor. Oil should be at normal (43° C – 110° F) operating temperature, then set engine to run at 3400-3450 rpm and observe volume of flow. Volume should be at least 15 L/min. (4 gpm). If volume is low, first check for plugged inlet screen (8—Fig. 66).

**113. HIGH RELIEF PRESSURE.** Clean area thoroughly around forward and reverse test ports (F and R—Fig. 70), then remove plug from test port and install test gauge as shown. Apply parking brake, block wheels and disengage front-wheel drive (if so equipped). **Tractor must be securely blocked, high range selected and brakes securely set when making this test. Even so, tractor may attempt to move. Be extremely careful.** Make sure fluid is at proper level, then start and run engine until fluid reaches normal (43° C – 110° F) operating temperature, then shift range lever to high range. Run engine at fast idle speed, depress control pedal to slow forward or reverse speed, observe gauge pressure, then let pedal return to neutral. **Check relief pressure for only a short time or overheating and damage may occur.** Remove test gauge, install plug, then install gauge in opposite port. Start engine, move control pedal in opposite direction and check pressure in opposite direction. Port (F—Fig. 70) is for checking forward pressure and port (R) is for checking pressure in reverse. Correct pressure is 41,370-44,817 kPa (6000-6500 psi). High pressure relief valves are shown at (40 and 56—Fig. 69).

**114. CONTROL LINKAGE ADJUSTMENT.** Raise and block rear axle so wheels are off ground and disengage front-wheel drive (if so equipped). Operate controls while checking linkage for freedom of movement or excessive wear. Linkage should operate smoothly and return to neutral. Lubricate grease fittings on shafts with lithium based multipurpose grease. Initial length of control rods from forward and reverse pedals to control levers is 346 mm (13.62 in.). One end of control rods has left hand thread, so that length can be changed without disconnecting ends by turning rod after loosening locknuts. Installed length of lower spring should be adjusted by turning nuts of eye bolt which attaches front of spring. Coil length of spring should be 133 mm (5.25 in.).

Start engine and operate at half speed. Move range shifter to LOW range position and operate ground speed control pedal several times through forward,

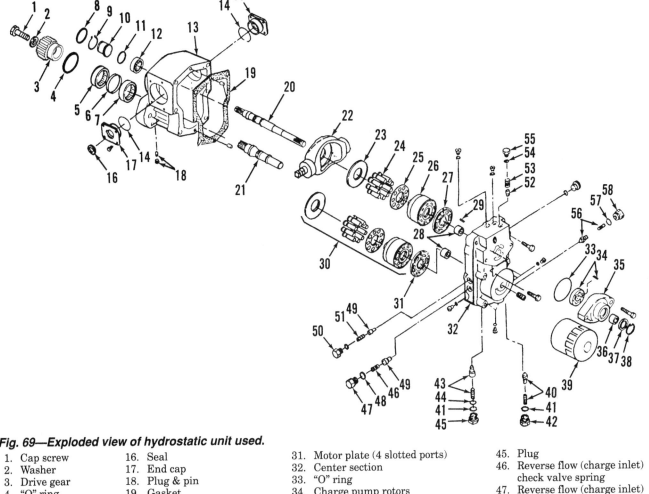

*Fig. 69—Exploded view of hydrostatic unit used.*

| | | |
|---|---|---|
| 1. Cap screw | 16. Seal | 45. Plug |
| 2. Washer | 17. End cap | 46. Reverse flow (charge inlet) |
| 3. Drive gear | 18. Plug & pin | check valve spring |
| 4. "O" ring | 19. Gasket | 47. Reverse flow (charge inlet) |
| 5. Ball bearing | 20. Top shaft | check valve plug |
| (same as 7) | 21. Motor shaft | 48. "O" rings |
| 6. Spacer | 22. Swashplate | 49. Valves |
| 7. Ball bearing | 23. Thrust plate | 50. Oil filter by-pass valve plug |
| (same as 5) | 24. Pump plungers | 51. Oil filter by-pass valve spring |
| 8. "O" ring | 25. Piston retainer | 52. Oil cooler by-pass valve |
| 9. Snap ring | 26. Cylinder block | 53. Oil cooler by-pass |
| 10. Seal guide | 27. Pump plate | valve spring |
| 11. "O" ring | (2 slotted ports) | 54. "O" ring |
| 12. Ball bearing | 28. Bearings | 55. Oil cooler by-pass valve plug |
| 13. Housing | 29. Pins | 56. Main relief valve & spring |
| 14. "O" rings | 30. Motor (similar to 23, | 57. "O" ring |
| 15. End cap | 24, 25 & 26) | 58. Plug |
| 31. Motor plate (4 slotted ports) | | |
| 32. Center section | | |
| 33. "O" ring | | |
| 34. Charge pump rotors | | |
| & drive pin | | |
| 35. Charge pump housing | | |
| 36. Needle bearing | | |
| 37. Seal | | |
| 38. Snap ring | | |
| 39. Canister filter | | |
| 40. Main relief valve & | | |
| spring | | |
| 41. "O" rings | | |
| 42. Plug | | |
| 43. Charge relief valve | | |
| & spring | | |
| 44. Shims | | |

neutral and reverse positions. Loosen cap screw (B— Fig. 71) and turn eccentric adjustment so that narrow part of eccentric is toward rear of tractor. Continue turning eccentric until the drive wheels stop turning, then find mid-point between where wheels begin turning forward and reverse. Hold eccentric at mid-point and tighten cap screw. In neutral, rear wheels should not creep either forward or reverse. Any movement of wheels indicates incorrect neutral adjustment. Refer to Fig. 72 for drawing of control linkage.

To adjust cruise control lock, remove platform shield, disengage master/park brake, disengage the cruise control lever, then depress reverse control pedal. Cruise control pawl (Fig. 72) must be 1-3 mm (0.04-0.12 in.) above ratchet. If clearance is incorrect, loosen screw which clamps position of the front eccentric, turn eccentric until pawl is correct distance above ratchet, then tighten clamp screw. Narrow side of eccentric should be toward front of tractor.

Make sure that master/park brake is disengaged, then depress forward control pedal fully and measure distance between pedal and footrest. Distance should be 7-10 mm (0.3-0.4 in.) above footrest. If measured distance is incorrect, loosen locknuts on ends of forward control rod, turn control rod until distance is correct, then tighten locknuts. One end of control rod

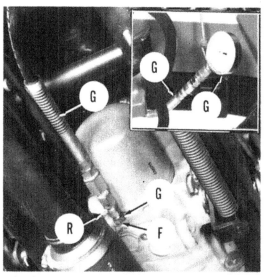
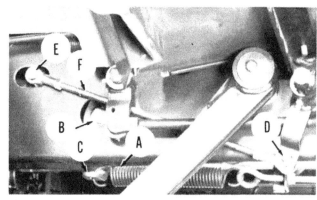

Fig. 71—Locking cap screw (B) and eccentric for adjusting neutral position are accessible through hole in right side frame.

A. Spring
B. Locking screw for eccentric
C. Neutral return lever

D. Locknut
E. Bolt
F. Speed control rod

*Fig. 70—To check high pressure relief pressure, attach gauge to port (F or R) as shown. If equipped with front-wheel drive, front drive shaft should be removed before attaching gauge to reverse port (R).*

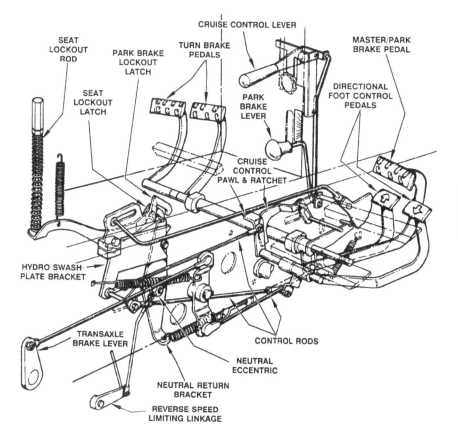

Fig. 72—Drawing of the hydrostatic controls. Eccentric adjustment for neutral is at pivot point of the neutral return bracket.

has left hand thread, so that length can be changed without disconnecting ends. Make sure that rod ends are parallel when locknuts are tight.

To check and adjust length of reverse control rod, make sure that master/park brake is disengaged and speed control pedals should be in neutral position. Pull upward on cruise control lever and check position of ratchet in relation to pawl. Ratchet must contact pawl at the first notch. If incorrect, loosen locknuts on ends of forward control rod, turn control rod until position is correct, then tighten locknuts. One end of control rod has left hand thread, so that length can be changed without disconnecting ends. Make sure that rod ends are parallel when locknuts are tight.

Distance between top of seat lockout rod (Fig. 72) and bottom of seat plate is 1-2 mm (0.04-0.08 in.) when seat is at rear and weight is removed from seat. Seat lockout latch should be tight against cam, preventing movement. Be sure that locknut is tightened if position of rod cap is changed.

## CHARGE PUMP

### All Models

**115. REMOVE AND REINSTALL.** To remove charge pump for the hydrostatic unit, first remove drive shaft as outlined in paragraph 24 (for 655, 755 and 756 models) or paragraph 55 (for 855, 856 and 955 models). Remove the two attaching screws and withdraw pump from input shaft.

Inner rotor is driven by pin. Front surface of hydrostatic housing should be smooth.

When installing charge pump, petroleum jelly can be used to hold drive pin in place while installing inner rotor. Position new "O" ring on pump housing and install outer rotor and pump housing with flat on side of pump housing toward right side of tractor. Pump will not operate if flat is on wrong side. Tighten the two retaining screws to 37-50 N·m (27-37 ft.-lbs.) torque. Refer to paragraph 24 for installing drive shaft in 655, 755 and 756 models. Refer to paragraph 55 for installing drive shaft in 855, 856 and 955 models.

## HYDROSTATIC UNIT

### All Models

**116. REMOVE AND REINSTALL.** To remove the hydrostatic unit, first remove drive shaft as outlined in paragraph 24 (for 655, 755 and 756 models) or paragraph 55 (for 855, 856 and 955 models). Refer to paragraph 115 and remove charge pump.

Disconnect battery ground cable from negative terminal of battery and remove floor panels. If equipped

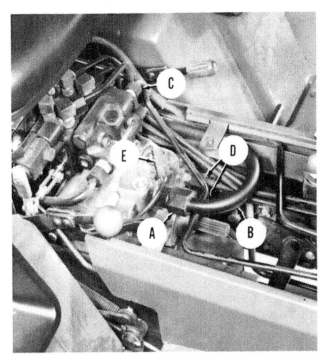

*Fig. 73—Detach suction line (B), steering line (C), oil cooler lines (D) and neutral start switch wire (E), before removing the hydrostatic transmission.*

with front-wheel drive, remove front-wheel drive shaft. For all models, refer to paragraph 109 and drain rear axle center housing. Disconnect spring (A—Fig. 71) from lever (C) and remove bolt (E) to detach rod (F). Detach fitting (A—Fig. 73), loosen clamp at rear end of line, then turn tube (B) up, out of the way. Detach steering line (C), cooler lines (D) and neutral start switch wire (E). Attach a lifting eye to top of hydrostatic unit and remove the two attaching screws. Move hydrostatic unit toward front until output gear is free of rear axle center housing, then lower unit to the floor.

Reinstall by reversing removal procedure. Be sure that coupling (seal) for upper shaft and "O" ring which seals the lower shaft are in place before tightening the two attaching screws to 142 N·m (105 ft.-lbs.) torque. Refer to paragraph 114 and adjust control linkage. Refer to paragraph 24 for installing drive shaft in 655, 755 and 756 models. Refer to paragraph 55 for installing drive shaft in 855, 856 and 955 models. Clean filter screen, install new cartridge filter, refill system reservoir with "John Deere Low Viscosity HY-GARD" transmission and hydraulic oil or equivalent, then bleed system as outlined in paragraph 109. Capacity is approximately 17 L (18 quarts), but some oil may remain in system. Control linkage should be lubricated with lithium base multipurpose grease.

**117. OVERHAUL** Before disassembling, plug all openings and clean exterior of hydrostatic unit. To disassemble, remove canister filter (39—Fig. 69) and charge pump (33-38). Remove cap screw (1), washer (2) and gear (3). Loosen all six screws which attach center housing (32) to housing (13), position hydrostatic unit on its side and carefully remove the six screws. **Be careful not to allow any internal parts to fall or otherwise become damaged while disassembling.** Hold motor shaft (21) and bump center housing (32) gently to separate center housing from housing (13). Carefully lift parts from housings. **Separate similar parts of pump and motor for correct assembly. Even though parts may be identical, parts should be reinstalled in original location.**

Valve plates (27 and 31) are located on pins (29). Carefully remove valve plates and inspect all parts of pump (23-27) and motor (30 and 31) for scoring. Renew any of these parts that are even lightly scored as indicated by light scratches. Do not remove bearings unless new bearings are to be installed. Parts (23 through 26) for pump may be similar to parts (30) for motor, but should not be interchanged. Thrust plate, for early production hydrostatic motor (30), is 1.02 mm (0.040 in.) larger in diameter than similar thrust plate (23) for pump. Do not attempt to use later, small diameter, thrust plate in early units which require the larger thrust plate for motor. Make sure that pistons operate freely in cylinder bores of pump and motor. Unbolt swashplate caps (15 and 17), then tap end of swashplate trunnion shaft (22) with a soft hammer to remove covers from case.

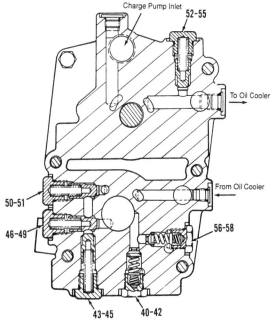

*Fig. 74—Cross section of hydrostatic unit center housing, showing positions of valves.*

| | |
|---|---|
| 40-42. Main relief valve | |
| 43-45. Charge relief valve | 50-51. Oil filter by-pass valve |
| 46-49. Reverse flow (charge inlet) check valve | 52-55. Oil cooler by-pass valve |
| | 56-58. Main relief valve |

Valves can be removed from center housing (32) for cleaning, but be sure to keep parts separate. Many valve parts (40-58) are similar, but should only be reinstalled in original location. Refer to Fig. 74 for cross section showing valves.

It is necessary to remove plug and pin (18—Fig. 69) to remove motor shaft (21) and bearings (5 and 7). Bearing retaining pin is threaded so that it can be pulled using a cap screw. The pin of early models is tapped 4-40 UNC and later models are tapped 8-32 UNC. It is necessary to remove snap ring (9) and seal guide (10) before removing bearing (12) and shaft (20). Use a knife edge puller to press bearings (5, 7 and 12) from shafts (20 and 21).

Check all parts for visible signs of wear or scoring and renew any that are cracked or otherwise questionable.

Clean all parts and housings thoroughly, then coat with new transmission oil before assembling. Coat "O" rings and seals with grease to provide initial lubrication and to hold parts in place while assembling.

Reassemble by reversing disassembly procedure. Lubricate swashplate (22) and end caps (15 and 17) with transmission oil, then position in housing (13). Be careful not to damage lip seal (16). Tighten screws retaining end caps (15 and 17) to 8-9 N·m (72-84 in.-lbs.) torque. Lubricate, then position thrust plates (23) for both pump and motor in housing (13) around the installed shafts (20 and 21). Thrust plate, for early production hydrostatic motor (30), is 1.02 mm (0.040 in.) larger in diameter than similar thrust plate (23) for pump. Do not attempt to use later, small diameter, thrust plate in early units which require the larger thrust plate for motor. Install pump and motor assemblies (24-26 and 30) over the appropriate shafts and against thrust plates (23).

Make sure that bearings (28) protrude approximately 3 mm ($7/64$ in.) above machined surface of center housing (32). Lubricate valve plates (27 and 31) and install over bearings and pins (29). Top (pump) plate (27) can be identified by having only two ports slotted. The lower (motor) plate (31) has four ports slotted. Refer to Fig. 75.

Position gasket (19—Fig. 69) over dowel pins and install center housing (32) over shafts (20 and 21). Install the six screws which attach housings together and tighten evenly, using a crossing pattern. Screws should not be tightened completely. Center housing may feel springy against springs of the cylinder assemblies until screws are completely tight. Gradually tighten screws evenly to a final torque of 44-55 N·m (33-41 ft.-lbs.), while turning shaft to check for correct and straight assembly. Lubricate seal on canister filter (39) and install filter.

Front surface of hydrostatic housing should be smooth. Install pin and charge pump inner rotor (34) on input shaft. Petroleum jelly can be used to hold

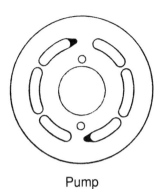

Pump

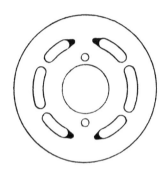

Motor

*Fig. 75—Valve plate for pump has two slotted ports as shown at dark areas and valve plate for motor has four.*

drive pin in place while installing inner rotor. Lubricate outer rotor and install in bore of housing (35). Position new "O" ring (33) on pump housing, then install outer rotor and pump housing with flat on side of pump housing toward right side of tractor. Pump will not operate if flat is on wrong side. Tighten the two retaining screws to 37-50 N·m (27-37 ft.-lbs.) torque.

## RANGE TRANSMISSION

### All Models

**118.** The two-speed range transmission provides a choice of two gear selections. The range transmission is located in front compartment of rear axle center housing and sliding shift coupling is located on bevel drive pinion shaft. Drive from the hydrostatic unit's lower shaft turns gear (3—Fig. 69 and Fig. 76), which then turns the High/Low intermediate gears and shaft (10, 12 and 13—Fig. 77). Shift coupling (24) can be moved to engage Low gear (21) with bevel pinion shaft (15) or to engage High gear (26) with bevel pinion shaft.

Mid pto drive, rear pto system and main hydraulic pump are also contained in or attached to the rear axle center housing. The pto and hydraulic pump are driven by upper shaft (60).

**119. REMOVE AND REINSTALL.** To remove rear axle center housing, remove hydrostatic unit as outlined in paragraph 116, rockshaft housing as outlined in paragraph 145, and both axle housings as outlined in paragraph 124. Unbolt and remove the Roll Over Protection Structure (ROPS), seat and platform. Detach or remove the linkage for front-wheel drive engagement, tractor brakes, pto control, range shifter control and rear differential lock. Remove drive shaft for front wheel drive and disconnect interfering wires from electrical components. Disconnect interfering lines from flow divider and selective control valves, then remove valves. Block front wheels to stop all movement, support front section of tractor and connect hoist to lifting eyes attached to rear axle center housing.

NOTE: Center housing may be damaged if supported improperly from below.

Remove cap screws attaching center housing to tractor frame and remove supporting cover from left side, then detach hydraulic line for pto from left side cover. Move center housing away from tractor and support safely so that unit can be disassembled.

When reinstalling, make sure that all parts are clean and attach lifting eyes to the center housing. Attach hoist to lifting eyes and carefully move housing into position against tractor frame. If axles are attached, center housing can be supported by jacks under axle housings, but housing may be damaged if improperly supported from below. Attach pto hydraulic line to left side cover using a new "O" ring and tighten fitting to 49 N·m (36 ft.-lbs.) torque. After pto hydraulic line is attached, install supporting cover to left side with the three M14 × 35 cap screws and install the remaining seven M14 × 35 attaching screws. Tighten all ten M14 × 35 screws to 142 N·m (105 ft.-lbs.) torque. Refer to paragraph 116 and install hydrostatic unit using new "O" ring seals. Refer to paragraph 124 and install axle housings if not already attached. Refer to paragraph 145 and install rockshaft housing. Install flow divider and selective control valves. Connect hydraulic lines, electrical wires and controls. Install ROPS and tighten retaining screws to 215 N·m (159 ft.-lbs.) torque. The one shorter screw on each side of ROPS should be installed with a thick flat washer and lock washer. The two longer screws on each side should be fitted with only a lock washer. Install rear wheels and tighten rear wheels lug bolts to 115 N·m (85 ft.-lbs.) torque. Refer to appropriate paragraphs for adjustment of control linkages.

**120. OVERHAUL.** If "O" ring seal (70—Fig. 77) is leaking, it can be renewed without removing center housing, if extreme caution is exercised. Drain fluid from center housing and disconnect control linkage

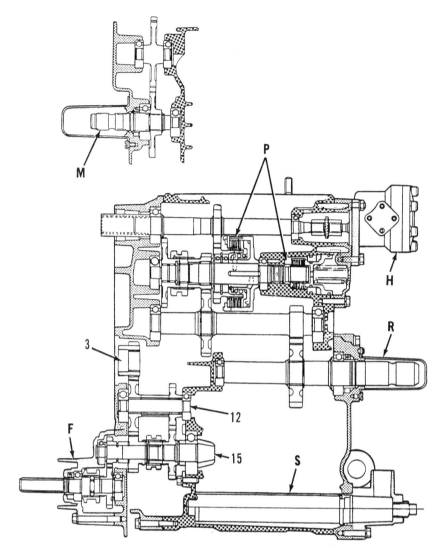

Fig. 76—Cross section of rear axle center housing showing some of the components installed inside or attached to the housing or covers.

3. Traction drive gear
12. Intermediate shaft
15. Bevel pinion
F. Front drive
H. Hydraulic pump
M. Mid pto
P. Pto clutch and brake
R. Rear pto
S. Hydraulic screen

from lever (64). Remove roll pin and lever (64), then remove screw and retainer plate (71). Shift transmission to Low range (turn shaft counterclockwise), then push shaft straight in, toward left side of tractor to disengage internal lever (66) from shift fork (67). **Mark outside of center housing where hole for roll pin which retains lever (64) is aligned, before doing any more twisting or pulling of shaft (65).** Carefully turn shaft clockwise so that internal lever (66) will be toward front without engaging or moving shift fork (67). When internal lever is disengaged from fork, shaft (65) can be pulled out (to the right) enough to remove and install "O" ring (70). Lubricate "O" ring and reinsert shaft, turn shaft until roll pin hole is aligned with mark on side of center housing, then pull shaft out (toward right) to engage internal lever with shift fork. Complete assembly by reversing disassembly procedure.

**121.** To remove range gears, shafts, bearings and related parts, first remove rear axle center housing as outlined in paragraph 119. If tractor is equipped with front-wheel drive, refer to paragraph 17 and remove front drive gear case. Remove the 17 screws attaching front cover (63—Fig. 77) of all models. Turn High/Low shifter shaft (65) as front cover is removed to allow shift fork (67) and rail (69) to remain in center housing. All gears and shafts should remain in center housing when separating.

If pto shift fork (P—Fig. 78) is removed, remove select switch pin until ready to assemble so that it will not be lost. The necessity to remove gears and shafts will depend upon service to be performed. Refer to paragraphs 131 through 135 for service to pto gears, shafts, clutch and brake assemblies.

High reduction gear (10—Fig. 77) has 31 teeth and high range gear (26) has 20 teeth for all models. Low reduction gear (13) has 23 teeth and Low range gear (21) has 28 teeth for 655, 755, 756, 855 and 856 models. Low reduction gear (13) has 21 teeth and Low range gear (21) has 30 teeth for 955 models. Front-wheel-drive gear (29) has 13 teeth on all models so equipped.

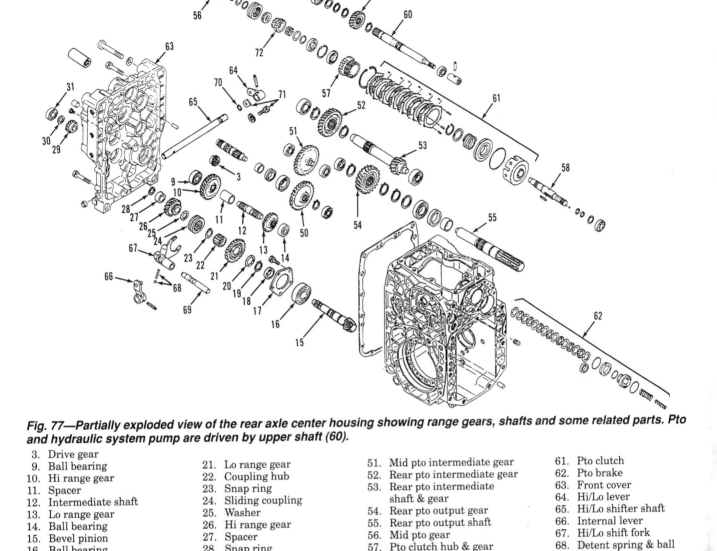

*Fig. 77—Partially exploded view of the rear axle center housing showing range gears, shafts and some related parts. Pto and hydraulic system pump are driven by upper shaft (60).*

| | | | |
|---|---|---|---|
| 3. Drive gear | 21. Lo range gear | 51. Mid pto intermediate gear | 61. Pto clutch |
| 9. Ball bearing | 22. Coupling hub | 52. Rear pto intermediate gear | 62. Pto brake |
| 10. Hi range gear | 23. Snap ring | 53. Rear pto intermediate | 63. Front cover |
| 11. Spacer | 24. Sliding coupling | shaft & gear | 64. Hi/Lo lever |
| 12. Intermediate shaft | 25. Washer | 54. Rear pto output gear | 65. Hi/Lo shifter shaft |
| 13. Lo range gear | 26. Hi range gear | 55. Rear pto output shaft | 66. Internal lever |
| 14. Ball bearing | 27. Spacer | 56. Mid pto gear | 67. Hi/Lo shift fork |
| 15. Bevel pinion | 28. Snap ring | 57. Pto clutch hub & gear | 68. Detent spring & ball |
| 16. Ball bearing | 29. Front-wheel drive gear | 58. Pto clutch & brake shaft | 69. Shift rail |
| 17. Retainer | 30. Washer | 59. Pto drive gear | 70. "O" ring seal |
| 18. Spacer | 31. Ball bearing | 60. Hydraulic pump | 71. Retainer plate & screw |
| 19. Snap ring | 50. Mid pto output gear | & pto drive shaft | 72. Rear pto gear |
| 20. Washer | | | |

On all models, screws attaching retainer (17) to center housing should be tightened to 26 N·m (19 ft.-lbs.) torque. Install washer (20) with grooved side out toward gear (21). Refer to Fig. 77 and Fig. 78 and assemble remaining parts as shown. Assemble detent (68), shift fork (67) and rail (69). Lubricate shift fork (67) and shift rail (69), then insert fork into shift collar (24) and rail into bore of center housing. If removed, assemble shifter shaft (65), "O" ring (70) and internal lever (66) in front cover (63), then install retainer (71). Install new front cover gasket on dowel pins, then lubricate all shafts and bearings to facili-

tate installation of front cover. Rotate shifter shaft (65) and internal lever (66) back so that pin of internal lever (66) can be engaged with groove in shift fork (67) as front cover is guided into position against rear axle center housing. The front cover is retained by four M12 screws which should be tightened to 90 N·m (66 ft.-lbs.) torque, six M10 screws which should be tightened to 50 N·m (37 ft.-lbs.) torque, and seven M8 screws which should be tightened to 26 N·m (19 ft.-lbs.) torque. Screws retaining front drive gearcase of models with front-wheel drive should be tightened to 26 N·m (19 ft.-lbs.) torque.

*Fig. 78—View of rear axle center housing with front cover removed showing shafts and gears in place.*

D. Differential lock shaft
P. Pto shift fork & shaft
10. Hi range gear
15. Bevel pinion
50. Mid pto output gear
51. Mid pto intermediate gear
52. Rear pto intermediate gear
56. Mid pto gear
60. Hydraulic pump & pto drive shaft

# DIFFERENTIAL

## DIFFERENTIAL LOCK AND SHIFTER

### All Models

**122.** The differential lock control rod can be adjusted to make sure that lock is engaged before pedal hits footrest. To check differential lock, start engine and allow to continue running at slow speed. Depress forward pedal to begin moving tractor, engage differential lock by pressing pedal and stop engine. The differential lock pedal should stay down (engaged), so that distance between bottom of pedal and footrest can be measured. The measured clearance between pedal and footrest should be 15 mm (0.6 in.) and can be adjusted by changing length of control rod.

The rear cover, axle housings, differential assembly, front cover, range transmission and bevel pinion shaft must first be removed to service the differential lock shaft (8—Fig. 79), fork (7), coupling (19). Refer to appropriate paragraphs for removal. Compress spring (5) and remove snap ring (3). A special tool can be fabricated from pipe with 25 mm (1 in.) inside diameter by grinding a 20 mm (¾ in.) notch in one end of the pipe. Special pipe tool should be long enough to slide over end of shaft (8) to compress the spring (5) while removing snap ring (3). Remove

washer (4) and spring (5). Work through hole in center housing and drive pin (6) from shaft, then withdraw shaft from right end.

Inspect all parts and renew any that are worn or damaged. Open side of larger roll pin (9) should be toward right end of shaft and smaller (inner) roll pin should be installed with open side 180° away from large pin (toward closed end of fork notch). Position shift fork (7) and install shaft (8) through bores in center housing and fork, then install roll pin (6) with closed sides of pin against fork. Install spring (5), washer (4) and snap ring (3), then use the unnotched end of special tool made from pipe with 25 mm (1 in.) inside diameter to slide over shaft (8) and compress spring until snap ring (3) enters its groove in shaft.

Refer to paragraph 123 for service to coupling (19), side gear (26) and remainder of differential.

## DIFFERENTIAL AND BEVEL GEARS

### All Models

**123. R&R AND OVERHAUL.** The differential can be removed and serviced without removal of bevel pinion (15—Fig. 79), but bevel gears (15 and 20) are

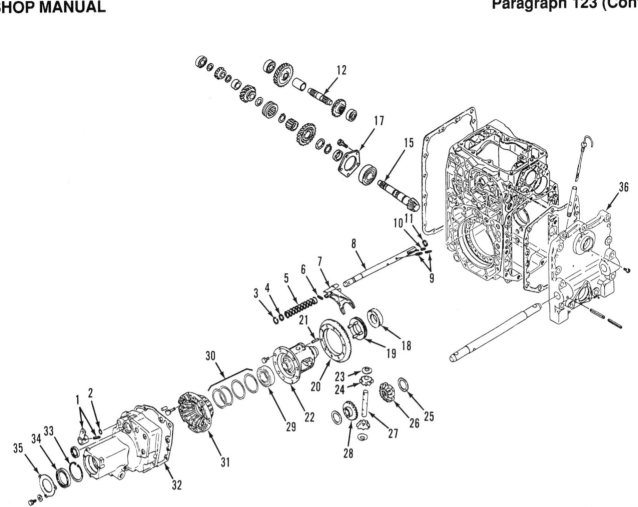

**Fig. 79—Differential is locked by sliding pins of collar through differential housing (22) and into notches in side gear (26). Parts (12, 15 & 17) are also shown in Fig. 77.**

1. Lever & pin
2. "O" ring
3. Snap ring
4. Spacer (same as 11)
5. Spring
6. Pin (same as 10)
7. Differential lock fork
8. Differential lock shaft
9. Roll pins

10. Roll pin (same as 6)
11. Spacer (same as 4)
12. Intermediate shaft
15. Bevel pinion
17. Retainer
18. Ball bearing
19. Locking collar
20. Ring gear

21. Pin
22. Differential housing
23. Thrust washer
24. Pinion
25. Thrust washer
26. Right side gear
27. Cross shaft
28. Left side gear

29. Ball bearing
30. Shims
31. Differential carrier
     housing
32. Axle housing
33. Snap ring
34. Seal
35. Retainer
36. Rear cover

available only as a matched set. If necessary to remove the bevel pinion, refer to paragraphs 119 and 120 for removal of the range transmission and bevel pinion (15).

To remove the differential assembly, first drain oil from the transmission and hydraulic system, then remove rear cover from the center housing. Remove axle assemblies, then unbolt and remove the carrier housing (31). Move differential assembly to the left and lift from center housing.

Shift collar (19) can be removed after pulling bearing (18) from shoulder of housing (22). Ring gear (20) can be removed after removing the ten retaining screws. Drive pin (21) through housing bore and remove cross shaft (27), pinions (24), thrust washers (23), side gears (26 and 28) and thrust washers (25).

Check all parts for wear, damage and freedom of movement.

Assemble differential using new spring pin (21). Side gear (26) on right side has notches for engagement of the differential lock (19). Clean threads in ring gear (20) and retaining screws with cleaner/primer and coat threads of retaining screws with medium strength thread lock. Tighten all ten screws evenly to 26 N·m (19 ft.-lbs.) torque. Make sure that ring gear is seated firmly and squarely against differential housing flange.

Shims (30) adjust backlash between ring gear (20) and bevel pinion (15). Install differential assembly (18 through 28) in center housing. Install bearing (29) and original or selected shims (30) in carrier housing. Coat edge of carrier housing (31), which pilots into bore of center housing, with grease before installing. Tighten screws retaining carrier housing evenly to 26 N·m (19 ft.-lbs.) torque.

To check backlash, push the installed differential assembly to left by tapping lightly against right side of differential housing (22) and bearing (18). The installed shims (30) should be tight between bore of carrier housing (31) and bearing (29). Left side bearing (29) should be tight against shoulder of differential housing (22). Attach a dial indicator to measure radial movement at outside edge of ring gear (20) teeth and measure backlash. A rod can be temporarily inserted through an opening in right side of center housing to hold bevel pinion from turning, while checking. Correct backlash is 0.15-0.21 mm (0.006-0.008 in.). Add or remove shims (30) as required to provide correct backlash. Shims are available in thickness of 0.1, 0.3 and 0.5 mm (0.004, 0.012 and 0.020 in.). Complete assembly by reversing disassembly. The four M12 × 70 screws and two M12 × 55 screws retaining rear cover should be tightened to 90 N·m (66 ft.-lbs.) torque. The five M10 × 50 rear cover retaining screws should be tightened to 50 N·m (37 ft.-lbs.) torque. The M10 screws retaining final drives should be tightened to 52 N·m (38 ft.-lbs.) torque. If not equipped with ROPS, the M16 screws retaining final drives should be tightened to 187 N·m (138 ft.-lbs.). If equipped with ROPS, the screws retaining ROPS and final drive should be tightened to 215 N·m (159 ft.-lbs.) torque. Rear wheel retaining lug bolts should be tightened to 115 N·m (85 ft.-lbs.) torque for all models.

# REAR AXLE, FINAL DRIVE AND BRAKES

## REAR AXLE AND FINAL DRIVE

### All Models

**124. REMOVE AND REINSTALL.** To remove the final drive and axle assembly, first drain fluid from transmission and hydraulic system as outlined in paragraph 109. Block tractor front wheels to prevent rolling, raise tractor and remove rear wheels.

**NOTE: Do not lift or support tractor under rear axle center housing. Weight of tractor may damage rear axle center housing if improperly lifted.**

Install jack stands under rear of tractor frame before removing rear wheels. Support ROPS, if so equipped, then unbolt and remove the Roll Over Protection Structure from tractor. Disconnect differential lock linkage from left side of all models, then remove pin and lever (1—Fig. 79). Disconnect brake control rods from both sides. Make sure that rear of tractor is supported securely, then support final drive and axle assembly safely and in a way that will allow unit to be withdrawn from center housing. If tractor is not equipped with ROPS, remove the three M16 retaining screws from upper rear part of axle housing. On all models, remove the six M10 retaining screws, then withdraw final drive and axle housing from center housing.

When installing, make sure that snap ring (51—Fig. 80) is in groove of input shaft and sun gear (50), then slide shaft through brake discs (62) and plates (63) and into splines of differential side gear (26 or 28—Fig. 79). Notch of separator plates (63—Fig. 80) should be aligned with bore in center housing for actuating cam shaft (40). Make sure alignment dowels, gasket and a new "O" ring (2—Fig. 79) are installed and properly positioned. Lubricate "O" ring (2), shaft (8) and bore in left axle housing. Position axle and final drive assembly against center housing, align teeth of planet gears (57—Fig. 80) with teeth of input shaft gear (50) and brake cam shaft (40) with notch of separator plates (63) and bore in center housing. When aligned, parts can slide together easily and the five M10 × 35 cap screws can be installed. One M10 × 50 cap screw is installed in top hole. Tighten all six M10 screws to 52 N·m (38 ft.-lbs.) torque. Install ROPS, if so equipped, and tighten retaining screws to 215 N·m (159 ft.-lbs.) torque. The one shorter screw on each side of ROPS should be installed with a thick flat washer and lock washer. The two longer screws on each side should be fitted with only a lock washer. For models without ROPS, install the six M16 screws and tighten to 187 N·m (138 ft.-lbs.) torque. Install rear wheels and tighten rear wheels lug bolts of all models to 115 N·m (85 ft.-lbs.) torque.

**125. OVERHAUL.** Planet assembly (B—Fig. 80) for 955 models is different from planet assembly (A) of other models. Refer to paragraph 124 for removal of axle and final drive for all models. Stand the axle on wheel flange end and detach brake return springs (59—Fig. 80). Carefully lift brake actuator disc (61) from axle, then remove the six actuator balls (60). A dial indicator can now be attached to housing (32) with pointer against planet carrier (48). Pull planet carrier (48) up and measure end play. End play can not be adjusted, and more end play than suggested maximum of 0.6 mm (0.024 in.) indicates worn parts. Remove snap ring (49) and lift planet reduction (46 through 52) from axle. If necessary, ring gear (44) can be unbolted and removed. It may be necessary to use

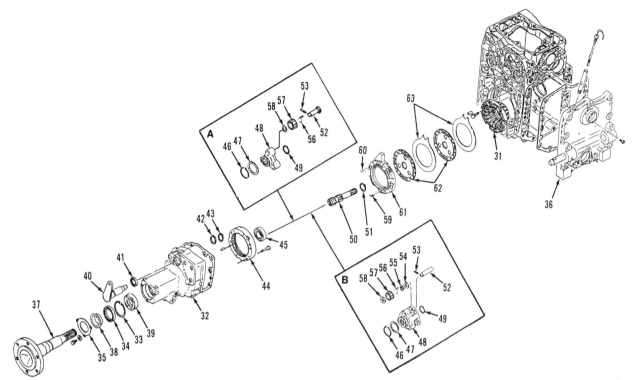

*Fig. 80—Exploded view of typical rear axle, final drive and brake for left side. Parts for right side are similar, but some such as brake lever and cam (40), actuator disc (61) and ring gear (44) are different. Final drive reduction (A) is used for all models except 955; reduction (B) is used on 955 models.*

| | | |
|---|---|---|
| 31. Differential carrier housing | 40. Brake lever & cam | 48. Planet carrier | 56. Needle rollers (22/path) |
| 32. Axle housing | 41. Seal | 49. Snap ring | 57. Planet gears (3/side) |
| 33. Snap ring | 42. Washer | 50. Input shaft & sun gear | 58. Washers |
| 34. Seal | 43. Snap ring | 51. Snap ring | 59. Brake return springs |
| 35. Retainer | 44. Ring gear | 52. Planet shafts (3/side) | (3/side) |
| 36. Rear cover | 45. Ball bearing | 53. Roll pins (3/side) | 60. Balls (6/side) |
| 37. Rear axle | 46. Snap ring | 54. Washers (same as 58) | 61. Actuator disc |
| 38. Seal sleeve | 47. Washer | 55. Washers | 62. Brake discs |
| 39. Ball bearing | | | 63. Separator plates |

two M8 jackscrews to push ring gear from the two alignment dowels.

The planet assembly can be disassembled by removing retaining pins (53) and pressing planet shafts (52) from carrier (48). Each planet gear (57) contains 22 loose needle rollers. Clean and inspect all parts for wear or visible damage.

When assembling planet assembly of 655, 755, 756, 855 and 856 models, observe the following. Set planet shafts (52) on bench with large end down, then position a planet gear (57) over shaft with beveled edge of teeth up. Coat the loose needle rollers (56) with grease and insert 22 rollers between each gear (57) and shaft (52). Install thrust washer (58) over shaft with grooved side against gear (57). After assembling pinion, bearing and shaft, press shaft into carrier (48) with notch in end of shaft (52) toward center of carrier (48) and holes for roll pin (53) aligned. Install roll pin with open side away from carrier (48). Bump shaft (52) down to make sure that roll pin is against carrier and planet gear is free to turn. After all planet gears

and shafts are installed in carrier, install thrust washer (47) and snap ring (46).

When assembling planet assembly of 955 models, observe the following. Set planet gears (57) on a clean surface with beveled edge of teeth up. Insert shaft (52) into gear, then insert washer (55) between shaft and gear. Coat the loose needle rollers (56) with grease and insert 22 rollers between each gear (57) and shaft (52). Install thrust washer (58) over shaft with grooved side against gear (57). Carefully lift gear, shaft and rollers from bench, install thrust washer (54) with grooves toward washer (55) and needle rollers, then slide shaft from the unit. Grease should hold loose needle rollers in position while installing. Position assembled gear and roller into carrier (48) with beveled end of teeth toward outer end of axle. After positioning pinion, bearing and thrust washers, press shaft (52) into carrier (48) with notch in end of shaft toward center of carrier (48) and holes for roll pin (53) aligned. Install roll pin with open side away from carrier (48). Bump shaft (52) down to make sure that roll pin is against carrier.

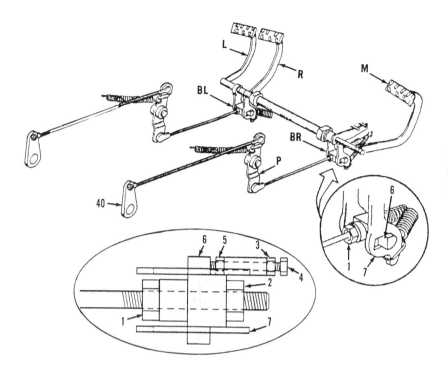

*Fig. 81—Refer to text for brake adjustment procedure. Adjuster blocks shown in the insets are located on both sides at (BL and BR).*

1. Locknut
2. Locknut
3. Jam nut
4. Adjusting screw
5. Jam nut
6. Adjuster block
7. Master/park pedal

After all planet gears and shafts are installed in carrier, install thrust washer (47) and snap ring (46).

Unbolt retainer (35), then bump axle shaft (37) from housing (32) and bearings (39 and 45). Outer bearing (39) can be pulled from housing after removing seal (34) and snap ring (33). Wear sleeve (38) can be pressed from axle.

Press bearings (39 and 45) into axle housing until tight against shoulder. Install snap ring (33), lubricate lip of seal (34), then press seal into bore until seated against snap ring. If new wear sleeve (38) is installed, be sure that retainer (35) is in place, lubricate axle and inside of wear sleeve, then press wear sleeve onto axle until inner edge is flush with shoulder of axle.

When installing axle, wrap splines with tape to reduce damage to lip of seal (34), then insert axle into housing and through bearings (39 and 45). Seat axle in bearings, then tighten screws attaching retainer (35) to 26 N•m (19 ft.-lbs.) torque. If removed, install seal (41), brake lever and cam (40), washer (42) and snap ring (43).

Install planet assembly and snap ring (49). Holes in bottom of ring gear (44) for attaching return springs (59) are not symmetrical. Attach ring gear so that a spring mounting hole is not next to brake camshaft (40) so that springs can be attached properly. Coat threads of screws attaching ring gear (44) with medium strength thread locker and tighten to 26 N•m (19 ft.-lbs.) torque.

Grease the six actuator balls (60) and position in pockets of ring gear (44). Position actuator plate (61) on top of actuator balls and attach the three springs (59) to holes in ring gear and actuator plate. The shorter hook should be attached to ring gear.

## BRAKES

### All Models

**126. ADJUST.** Left, right and master/park brake should be checked and adjusted together in the following sequence.

Lock the master/park brake pedal (M—Fig. 81) in the first notch of park brake ratchet. Loosen locknuts (1) at the rear of adjuster blocks (BL and BR) on both sides, then loosen both locknuts (2) at front. Loosen front jam nuts (3), then loosen adjusters (4) until threads are flush with back of rear nuts (5).

Pry lower arm of bellcrank (P) toward rear with a screwdriver using enough pressure to just overcome tension of return spring and to engage brake, then turn rear nut (1) until the front of adjuster block (6) just contacts front of slot in the master/park pedal (7). Turn front locknut (2) against front of adjuster block (6), then tighten rear nut (1). After adjusting nuts (1 and 2) for right side, adjust left side using the same procedure.

After adjusting nuts (1 and 2) on both sides, turn adjusting screws (4) until adjuster block is square with pedal arms. Press both brake pedals with the same pressure. If both pedals are not even, turn adjuster screw (4) for left side as required to align pedals, then tighten locknuts (3) for both sides.

**127. R&R AND OVERHAUL.** Brake discs (62—Fig. 80) and separator plates (63) can be renewed after removing rear axle and final drive housing as outlined in paragraph 124. Stand axle and final drive assembly on wheel flange, use needle nose pliers to disconnect the three return springs (59), then actuator disc (61) can be removed. The six actuator balls (60) should be lifted from ramp pockets in face of ring gear (44), before they are lost. If brakes for both sides are disassembled, keep parts separate. Brake lever and cam (40), actuator disc (61), and ring gear (44) for left side are different than for similar parts for right side.

Brake discs (62) are 4.6-4.8 mm (0.181-0.189 in.) thick when new. New discs should be installed if thickness is less than 4.4 mm (0.173 in.). Separator plates (63) are 2.5-2.7 mm (0.098-0.106 in.) thick when new. New plates should be installed if thickness is less than 2.3 mm (0.090 in.) or if warped more than 0.3 mm (0.012 in.).

If ring gear with brake actuator ramps is removed, note that holes in bottom of ring gear (44) for attaching return springs (59) are not symmetrical. Attach ring gear so that a spring mounting hole is not next to brake camshaft (40) so that springs can be attached

properly. Coat threads of screws attaching ring gear (44) with medium strength thread locker and tighten to 26 N·m (19 ft.-lbs.) torque.

Grease the six actuator balls (60) and position in pockets of ring gear (44). Position actuator plate (61) on top of actuator balls and attach the three springs (59) to holes in ring gear and actuator plate. The shorter hook should be attached to ring gear.

Make sure that snap ring (51) is in groove of input shaft and sun gear (50), then slide shaft through brake discs (62) and plates (63), and into splines of differential side gear (26 or 28—Fig. 79). Notch of separator plates (63—Fig. 80) should be aligned with bore in center housing for actuating cam shaft (40). Make sure alignment dowels, gasket and a new "O" ring (2—Fig. 79) are installed and properly positioned. Lubricate "O" ring (2), shaft (8) and bore in left axle housing.

Refer to paragraph 124 for attaching final drive and axle to center housing.

# POWER TAKE OFF

**128.** The pto system consists of the mechanical components necessary to drive both mid and rear output shafts (55—Fig. 82 and Fig. 83), the hydraulic components necessary to control engagement of either pto clutch or brake, and the necessary mechanical and electrical components required to safely control operation. The pto clutch is engaged by applying hydraulic pressure behind operating piston (118—Fig. 84) to compress release spring (117) and to press clutch discs and plates (112) together. The pto brake is released by applying hydraulic pressure to piston (130) to compress springs (132 and 133), permitting brake discs and plates (62) to separate. Relief valve (1-13) delays engagement of the clutch and release of the brake until pressure is sufficient for proper engagement. Electrical switches are used to assist safe starting of the engine, safe engagement of the pto and to illuminate dash indicators.

## TESTS AND ADJUSTMENTS

### All Models

**129. ELECTRICAL TESTS.** Safety switches are installed in the interest of operator safety and should not be shorted or by-passed. First test operation of the key switch as follows. Remove 10 amp fuse from second position from top of fuse panel. Fuse location is indicated by pto symbol. Attach test light or meter to left side of fuse holder and turn key switch to "ON" position. Meter should indicate battery voltage or test light should be illuminated. Briefly turn key to

"START" position and light should go out, current is interrupted. Refer to wiring diagram (Fig. 61, Fig. 62, Fig. 63 or Fig. 64) for assistance in tracing fault.

If preceding test of key switch operated satisfactorily, replace pto fuse and continue test as follows. Tilt seat forward and turn key switch to "OFF" position. Separate the three pin connector for seat switch and check for continuity across switch terminals at connector. With seat switch not depressed, current should not pass between switch leads. With seat switch depressed, current should pass between pink/white wire and yellow wire. Current should pass between blue and yellow wires when seat switch is pulled up in the override position.

If preceding tests indicate that key switch and seat switch are functioning properly, reconnect seat switch connector and test time delay control (TDC) module as follows. Check black wire to switch (17—Fig. 84) for proper grounding and check switch for proper operation. Make sure that wire terminals are making good contact with terminals of switch (17). Detach wire connector to pto lever magnet, turn key switch "ON," then attach test light or meter to blue wires and check for battery voltage. Battery voltage at connector should cause magnet to click. If voltage is not available to blue wires of connector for pto lever magnet, check for battery voltage to blue wire at switch (17). If voltage is available at blue wire to switch (17), but not at blue wire connector for pto lever magnet, check condition of connecting wires. If blue wire connector at switch (17) does not have

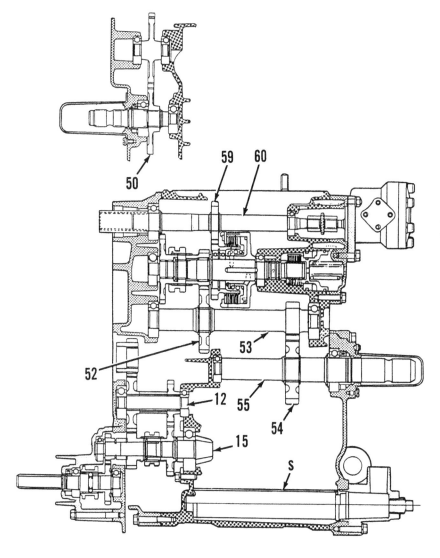

*Fig. 82—Cross section showing pto drive and related parts.*

12. Intermediate shaft
15. Bevel pinion
50. Mid pto output gear
52. Rear pto intermediate gear
53. Rear pto intermediate shaft & gear
54. Rear pto output gear
55. Rear pto output shaft
59. Pto drive gear
60. Hydraulic pump & pto drive shaft

battery voltage, check or renew time delay control (TDC) module.

**130. PTO CLUTCH ENGAGEMENT PRESSURE.** Remove plug from test port (P—Fig. 84) and attach a suitable pressure gauge. Make sure that oil is at normal temperature and run engine at 3450 rpm. Engage pto clutch and observe pressure. Pressure should increase from 207 to 827 kPa (30 to 120 psi) within 1.6 seconds for 655, 755, 756, 855 and 856 models. Pressure should increase from 227 to 1089 kPa (33 to 158 psi) within 1.6 seconds for 955 models. If pressure is too low, remove upper valve as outlined in paragraph 133 and add shims (6—Fig. 84).

## REAR OUTPUT SHAFT

### All Models

**131. REMOVE AND REINSTALL.** To remove rear cover (36—Fig. 83) and output shaft (55), first drain fluid from transmission and hydraulic system, then remove hydraulic fluid strainer as outlined in paragraph 109. Remove retaining screws, then pull rear cover from locating dowels.

Shaft (55) and bearing (77) can be bumped from rear cover, then disassembled as necessary. Inspect gear (54), bearings (73 and 77), seal (78), wear ring (79) and related parts for wear or damage. Wear sleeve (79) should be installed from rear (splined) end of output shaft (55) with chamfer toward bearing (77). Clean old gasket from surfaces of center housing and rear cover. Grease seal (78) and press into rear cover bore with spring loaded lip toward inside.

The four M12 × 70 screws and two M12 × 55 screws retaining rear cover should be tightened to 90 N·m (66 ft.-lbs.) torque. The five M10 × 50 rear cover retaining screws should be tightened to 50 N·m (37 ft.-lbs.) torque.

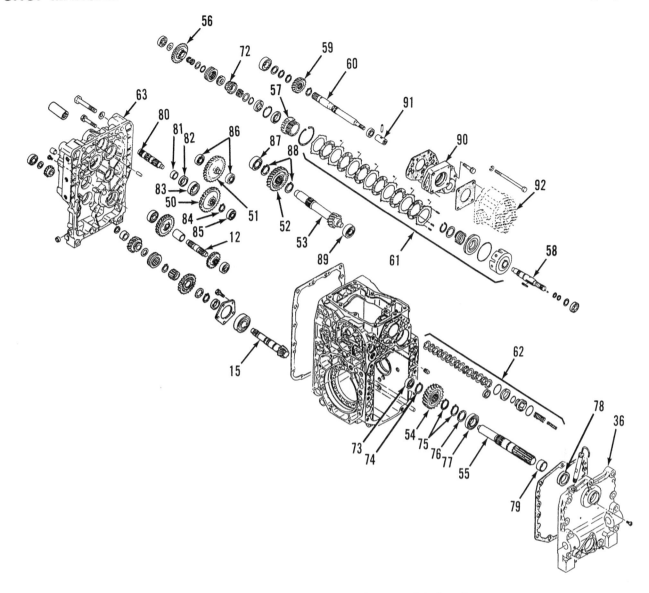

***Fig. 83—Exploded view of rear axle center housing, pto drive and some related parts.***

| | | |
|---|---|---|
| 12. Intermediate shaft | 56. Mid pto gear | 73. Ball bearing |
| 15. Bevel pinion | 57. Pto clutch hub & gear | 74. Snap ring |
| 36. Rear cover | 58. Pto clutch & brake shaft | 75. Snap rings |
| 50. Mid pto output gear | 59. Pto drive gear | 76. Washer |
| 51. Mid pto intermediate gear | 60. Hydraulic pump | 77. Ball bearing |
| 52. Rear pto intermediate gear |    & pto drive shaft | 78. Seal |
| 53. Rear pto intermediate shaft | 61. Pto clutch | 79. Wear ring |
|    & gear | 62. Pto brake | 80. Mid pto shaft |
| 54. Rear pto output gear | 63. Front cover | 81. Wear ring |
| 55. Rear pto output shaft | 72. Rear pto gear | 82. Seal |

| |
|---|
| 83. Ball bearing |
| 84. Snap ring |
| 85. Ball bearing |
| 86. Ball bearings |
| 87. Ball bearing |
| 88. Snap rings |
| 89. Bearing |
| 90. Cover and adapter |
| 91. Coupling |
| 92. Hydraulic pump |

## PTO INERTIA BRAKE

### All Models

**132. R&R AND OVERHAUL.** To remove pto inertia brake, first unbolt and remove hydraulic pump (92—Fig. 83) as outlined in paragraph 144. Remove coupling (91), then unbolt and remove cover (90).

**NOTE: The cover (90) is spring loaded, so cover retaining screws should be loosened evenly to prevent warping cover and possibly causing damage.**

Remove springs (132 and 133—Fig. 84), then use an 8M screw to pull piston (130) from bore.

Measure distance between rear gasket surface of rear axle center housing and back of last brake plate (62). If measured depth is more than 43 mm (1.693 in.), brake discs and plates may be worn. Each brake plate should be 1.0 mm (0.039 in.) thick and each brake disc should be 1.9 mm (0.075 in.) thick. Models 655, 755, 756, 855 and 856 models contain six discs and seven plates. Pto brake for 955 models contains seven discs and eight plates.

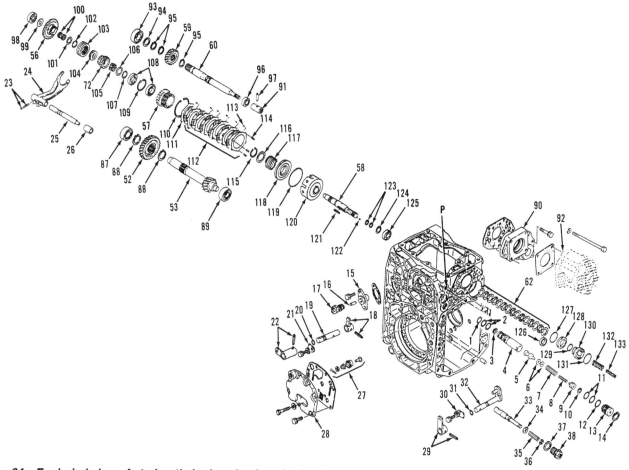

*Fig. 84—Exploded view of pto inertia brake, clutch and related parts.*

| | | | | |
|---|---|---|---|---|
| 1. Seal ring | 25. Shift rail | 62. Brake discs & plates | 110. Snap ring |
| 2. Seal rings | 26. Bushing | 72. Rear pto gear | 111. Top plate |
| 3. Seal ring | 27. Lube relief valve | 87. Ball bearing | 112. Lined discs |
| 4. Delay relief (upper) | 28. Cover | 88. Snap rings | & separator plates |
|    valve body | 29. Pto clutch/brake | 89. Bearing | 113. Separator springs |
| 5. Delay relief valve plunger |    outer lever | 90. Cover and adapter | 114. Pins (3 used) |
| 6. Shims | 30. Retainer plate | 91. Coupling | 115. Snap ring |
| 7. Spring | 31. "O" ring | 92. Hydraulic pump | 116. Washer |
| 8. Spring | 32. Control shaft & lever | 93. Ball bearing | 117. Release spring |
| 9. Cap | 33. Valve spool | 94. Washer | 118. Piston |
| 10. Seal ring | 34. Washer | 95. Snap rings | 119. "O" ring |
| 11. Seal rings | 35. Spring | 96. Ball bearing | 120. Clutch drum |
| 12. Seal ring | 36. Snap ring | 97. Pin | 121. Drive key |
| 13. Stop | 37. Gasket | 98. Ball bearing | 122. Plug |
| 14. Snap ring | 38. Switch & adapter | 99. Washer | 123. Fiber seal rings (2 used) |
| 15. Flange adapter | 52. Rear pto inter- | 100. Needle bearings | 124. Washer |
| 16. Select switch pin |    mediate gear | 101. Grooved washer | 125. Ball bearing |
| 17. Select switch | 53. Rear pto inter- | 102. Snap ring | 126. Bushing |
| 18. Internal select lever & pin |    mediate shaft & gear | 103. Shift coupling | 127. "O" ring |
| 19. Select shaft | 56. Mid pto gear | 104. Splined collar | 128. Bushing |
| 20. "O" ring | 57. Pto clutch hub & gear | 105. Roller bearing | 129. "O" ring |
| 21. Retainer plate | 58. Pto clutch & brake shaft | 106. Grooved washer | 130. Pto brake piston |
| 22. Outer select lever & pin | 59. Pto drive gear | 107. Snap ring | 131. "O" ring |
| 23. Detent ball & spring | 60. Hydraulic pump | 108. Ball bearings | 132. Spring |
| 24. Mid or rear pto shift fork |    & pto drive shaft | 109. Snap ring | 133. Spring |

Free length of larger spring (132) should be 64.3 mm (2.531 in.) for 655, 755, 756, 855 and 856 models. Spring for these models should exert 700.5 N (157.5 lbs.) force when compressed to 47.5 mm (1.870 in.).

Free length of larger spring (132) for 955 models should be 60.5 mm (2.382 in.) and spring should exert 1117 N (251.5 lbs.) force when compressed to 44.5 mm (1.75 in.).

Free length of smaller, inner spring (133) should be 61.6 mm (2.425 in.) for all models. Inner spring should exert 327 N (73 lbs.) force when compressed to 47.5 mm (1.870 in.).

Reinstall by reversing removal procedure, making sure that correct number of discs and plates (62) is installed. Install and coat new seal rings (127, 129 and 131) with grease before inserting into bore. Install springs (132 and 133), gasket and cover (90). Tighten the five cover retaining screws evenly to 26 N·m (19 ft.-lbs.) torque, then install coupling (91) and hydraulic pump (92). Tighten pump retaining stud nuts and screws to 26 N·m (19 ft.-lbs.) torque.

## CONTROL VALVES

### All Models

**133. R&R AND OVERHAUL.** To remove pto clutch and brake hydraulic control valves (1-14 and 33-38—Fig. 84), first disconnect external control linkage from lever (29), then remove lever and pin (29). Unbolt and remove cover (28) and withdraw lube relief valve assembly (27). Remove screw attaching retainer plate (30), then push shaft and lever (32) straight in to disengage shift lever from valve (33). Remove switch and adapter (38), then pull valve assembly (33-36) from bore. Remove snap ring (14), thread a 6M screw into stop (13), then pull stop and upper valve from bore.

Disassemble, install new seals and reassemble. Shims (6) adjust pto clutch engagement pressure as tested in paragraph 130. Complete assembly by reversing removal procedure.

## CLUTCH AND DRIVE SHAFTS

### All Models

**134. R&R AND OVERHAUL.** The pto clutch shafts (53, 58 and 60—Fig. 84) and related parts are removed together as follows. Refer to paragraph 119 and separate rear axle center housing from tractor. Refer to paragraph 132 and remove pto inertia brake, then refer to paragraph 121 and remove center housing front cover. Remove shift fork (24—Fig. 84), being careful not to lose detent ball and spring (23). The hydraulic pump and pto drive shaft (60), pto clutch and brake shaft (58), and the rear pto intermediate

shaft and gear (53) can be removed from rear axle center housing together with related gears and other parts. The mid pto intermediate gear (51—Fig. 83), mid pto gear (50) and mid pto output shaft (80) can also be removed.

To disassemble the clutch assembly, first use a knife edge puller to remove bearing (98—Fig. 84). Remove grooved washer (99), gear (56), needle bearings (100), washer (101) and coupling (103). Remove snap ring (102), splined collar (104), gear (72), roller bearing (105) and washer (106). Remove snap ring (107), then withdraw hub and gear (57) with bearings (108) and snap ring (109). Slide shaft (58) from clutch assembly and remove key (121).

Check general condition of discs and plates before disassembling clutch. Push the last plate, next to piston (118), toward snap ring (110) and measure distance between tab of plate and bottom of slot in drum (120). With separator springs (113) compressed, measured distance should be less than 4.7 mm (0.185 in.). If distance is excessive, install new discs and plates (112).

Remove snap ring (110) and top plate (111), then measure thickness of top plate. Install new top plate (111) if less then 2.9 mm (0.114 in.) thick. Remove the 15 or 18 small separator springs (113) as clutch discs and plates (112) are lifted from drum (120). Models 655, 755, 756, 855 and 856 are equipped with five clutch discs, six clutch plates and 15 separator springs (113). Model 955 is equipped with six clutch discs, seven clutch plates and 18 separator springs (113). On all models, measure thickness of clutch discs and plates. Install new clutch discs if worn to less than 1.9 mm (0.075 in.) and separator plates if worn to less than 1.0 mm (0.039 in.).

A special tool can be fabricated to press against washer (116), compressing spring (117) so that snap ring (115) can be removed, then washer (116), spring (117) and piston (118) can be removed. Free length of release spring (117) should be at least 29 mm (1.14 in.) and spring should exert at least 540 N (121 lbs.) force when compressed to 17.5 mm (0.689 in.).

Clean and inspect all parts for wear. Remove seal ring and install new ring being careful not to twist ring in groove. Lubricate seal (119) and piston (118), then insert piston in drum (120). Position spring (117) and washer (116), then use special fabricated tool to compress spring so that snap ring (115) can be installed. Install one flat separator plate in bottom of the drum, then position the three pins (114) in every other hole in lugs. Install one spring (113) over each pin, then install one clutch disc followed by another plate. Install three more separator springs (113) over pins, one clutch disc and plate. Continue assembling until all clutch discs, plates and separator springs are positioned, then install thicker top plate (111) and snap ring (110).

Assemble clutch and brake shaft by reversing disassembly procedure. Install washers (106, 101 and 99) with lubrication grooves toward needle or roller bearings (100 and 105). Press ball bearing (98) onto shaft and check for free movement of shift collar (103) and gears.

Gear (59) has 19 teeth for 655, 755, 756, 855 and 856 models; 17 teeth for 955 models. Gear and hub (57) has 28 teeth on 655, 755, 756, 855 and 856 models; 25 teeth on 955 models. Gear (52) has 35 teeth on all models. Bearing (89) is ball type for 655, 755, 756, 855 and 856 models. On 955 models, bearing (89) has a thrust flange on inner race and flange should be toward the 11 tooth gear of shaft (53).

Install 30 tooth gear (50—Fig. 83) on mid pto shaft with hub toward rear, snap ring groove, end of shaft (80). Coat inside of oil seal wear sleeve (81) with grease and press onto shaft with chamfered end against bearing (83). Press wear sleeve (81), bearing (83) and gear (50) tight against snap ring (84). Bearings (86) on ends of intermediate shaft and gear (51) are identical.

All three shafts (53, 58 and 60—Fig. 84), clutch, shift fork (24) and rail (25) should be installed together. Hold clutch and shaft (58) in place, position shift rail and fork (23-25), shafts (53 and 60) and

related parts, then slide all shafts into bores of rear axle center housing at the same time. Refer to paragraph 119 and install front cover. Refer to paragraph 132 and install pto inertia brake assembly.

## MID PTO

### All Models

**135. R&R AND OVERHAUL.** The mid pto output shaft (80—Fig. 83), intermediate gear (51) and related parts can be removed and serviced after removing the front cover from rear axle center housing. Refer to paragraph 119 and separate rear axle center housing from tractor, then refer to paragraph 121 and remove center housing front cover.

Install 30 tooth gear (50) on mid pto output shaft with hub toward rear, snap ring groove, end of shaft (80). Coat inside of oil seal wear sleeve (81) with grease and press onto shaft with chamfered end against bearing (83). Press wear sleeve (81), bearing (83) and gear (50) tight against snap ring (84). Bearings (86) on ends of intermediate shaft and gear (51) are identical.

# HYDRAULIC LIFT SYSTEM

**136.** The hydraulic lift pump (92—Fig. 84) is mounted on rear of the rear axle center housing and is driven by shaft (60), at engine speed, whenever the engine is running. The pump supplies pressurized fluid for steering, rockshaft control and remote control valves. Pressurized fluid flows from pump to a proportional flow divider valve, which first directs fluid to the steering system, before supplying lift systems.

## TROUBLESHOOTING

### All Models

**137.** The following are symptoms and possible causes which may occur during operation of the hydraulic systems. Refer to additional paragraphs for testing, adjusting and servicing.

1. Pump noise (cavitation) could be caused by:
   a. Hydraulic fluid low. Check and add fluid.
   b. Improper type or viscosity of hydraulic fluid.
   c. Suction pipe fittings leaking air.
   d. Reservoir screen and/or vent plugged.
   e. Hydraulic pump seal leaking.
   f. Pump mounting screws loose.

   g. Hydraulic pump scored or worn.
   h. Pump drive coupling worn or damaged.
   i. Hydraulic lines not clamped properly.

2. No hydraulic functions (no steering) could be caused by:
   a. Low hydraulic fluid level or improper oil.
   b. Air leak in suction line. Also check suction line to hydrostatic transmission charge pump.
   c. Leaking hydraulic pump seals.
   d. Plugged reservoir screen and/or vent.
   e. Broken or missing pump coupling.
   f. Hydraulic pump worn or damaged.
   g. System relief valve stuck open.
   h. Cracked flow divider housing.

3. No hydraulic functions (steering works or is slow) could be caused by:
   a. Same causes as 2.
   b. Rockshaft piston cover cracked.
   c. Hydraulic oil diverter plug installed in tractors without selective control valve. Plug is used only when selective control valve is installed.
   d. Mesh screen between rockshaft control valve and rockshaft housing plugged.
   e. "O" ring between rockshaft control valve and rockshaft housing leaking or blown.

f. Unloading valve stuck open.

g. Flow control valve stuck open.

h. Protective plugs left in ports of selective control valve. Plugs are to seal openings to prevent the entrance of dirt.

4. Slow operation could be caused by:

a. Same causes as 2.

b. Rate-of-Drop/Stop valve turned in.

c. System relief valve set too low.

d. Proportional flow divider valve installed incorrectly.

e. Mesh screen between rockshaft control valve and rockshaft housing plugged.

f. Unload valve or flow control valve leaking or malfunctioning.

g. Rockshaft spool valve scored.

5. Hydraulic oil overheating could be caused by:

a. Hydraulic system operated extensively in relief.

b. Operating with Stop valve closed and rockshaft control lever rearward.

c. Stop valve stuck closed.

d. Improperly adjusted feedback linkage. System in relief because control valve not returning to neutral.

e. Selective control valves not returning to neutral.

f. System relief set too low. Refer to paragraph 150.

g. Leak in rockshaft cylinder or implement relief valve.

h. Implement relief valve set too low or leaking. Refer to paragraph 151.

i. Transmission oil cooler plugged.

j. Transmission operated extensively in relief.

k. Oil cooler bypass valve stuck open.

l. Transmission closed loop relief valve set too low.

m. Tractor overloaded.

n. Pinched or restricted hydraulic line.

o. Improperly matched hydraulic component or implement.

6. Rockshaft hitch will not lift could be caused by:

a. Same causes as 2.

b. Stop valve closed.

c. System relief valve set too low. Refer to paragraph 150.

d. Load is too heavy.

e. Rockshaft piston scored or seals leaking.

f. Mesh screen between rockshaft control valve and rockshaft housing plugged.

g. Unloading valve stuck open.

h. Rockshaft spool valve scored or seals leaking.

i. Flow control valve stuck open.

j. Lower valve stuck open.

k. Lever linkage is bent.

7. Selective control valves fail to work or operate erratically could be caused by:

a. Same causes as 2.

b. Protective plugs left in ports of selective control valve. Plugs are to seal openings to prevent the entrance of dirt.

c. Hydraulic oil diverter plug not installed. Plug must be installed when selective control valve is installed.

## TESTS AND ADJUSTMENTS

### All Models

**138.** The selective control lever (S—Fig. 85) moves one valve when moved along axis of tractor and another valve when moved from side to side. Moving lever to right and forward usually extends the controlled cylinders.

To test, pull lever (S) to rear and release. Lever must return to "NEUTRAL" (center) position. Push lever forward to the first stop position and release. The lever must return to center position. Push lever fully forward to "FLOAT" position and release. The lever must stay in the "FLOAT" position.

Pull selective control lever (S) to the left (toward seat) and release. Lever must return to "NEUTRAL" (center) position. Push lever to the right (away from seat) to the first stop position and release. The lever must return to center position. Push lever fully to the right (the "FLOAT" position), and release. The lever must stay in the "FLOAT" position.

**139.** Check the rockshaft controls as follows: Turn rockshaft drop/lock valve (K—Fig. 86) counterclockwise as far as possible to open valve fully, then run engine at fast idle speed. Push rockshaft control lever (R—Fig. 87) fully forward to lower the rockshaft. It may be necessary to push rockshaft arms down to fully lower the arms. Pull rockshaft control lever (R) fully to the rear and make sure that rockshaft arms raise fully. With arms raised, pull the rockshaft lift arms up, then push them down to determine amount of free play. If the arms are not raised fully or if there is no noticeable free play, loosen locknut and adjust length of turnbuckle (T—Fig. 88). Shorten control rod

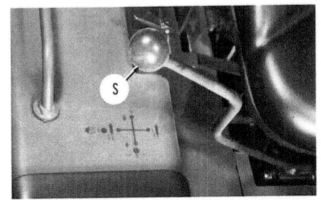

*Fig. 85—The selective control valves are controlled by moving lever (S).*

by turning the turnbuckle (T) until lift arms (A) are fully raised, then lengthen rod by turning the turnbuckle until valve just closes as determined by load on engine being noticeably reduced. From this (just fully raised) position, lengthen rod by turning the turnbuckle (T) an additional four to six flats and lock

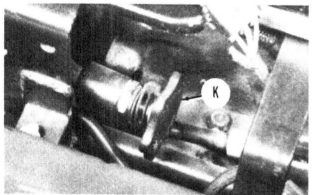

*Fig. 86—View of knob (K) for the drop/lock valve which is used to adjust (control) how fast the rockshaft lift arms fall. Valve must be adjusted to control speed of drop depending upon weight of load on arms and desired speed of drop.*

*Fig. 87—View of the rockshaft control lever (R) and adjustable depth stop (D).*

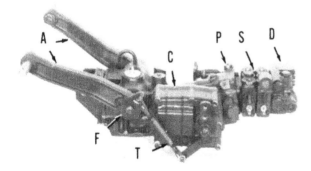

*Fig. 88—View of rockshaft housings control valves and related parts showing relative position.*

A. Rockshaft lift arms
C. Control valve cover
D. Flow divider valve
F. Feedback arm
P. Piston cover
S. Selective control valves
T. Turnbuckle

adjustment by tightening locknut. Check adjustment by lowering and raising lift arms, then recheck for some free play.

**140.** To check rockshaft drop/lock valve, move rockshaft control lever (R—Fig. 87) to raise the rockshaft arms fully and turn knob (K—Fig. 86) clockwise fully to close valve. Move rockshaft control lever forward to the lower position and notice lift arms. Rockshaft lift arms must not move. If arms move, repair the drop/lock (stop) valve.

Make sure that rockshaft control linkage is correctly adjusted as outlined in paragraph 139. Make sure that oil is at normal operating temperature, then raise lift arms and attach an implement or weight box that weighs 350 kg (770 lbs.) to lift arms. Stop engine and measure the amount of drop at end of lift arms (A—Fig. 88). Arms should not lower more than 2 mm (0.079 in.) in one minute. If drop is within limits, system is normal. If drop is more than 2 mm (0.079 in.), raise lift arms, close drop/lock valve (K—Fig. 86), stop engine and recheck time for weighted rockshaft arm to drop 2 mm (0.079 in.).

If weighted lift arms lower too fast with drop/lock valve closed, check for: 1. Leaking implement relief valve. 2. Rockshaft piston "O" rings leaking. 3. Scored rockshaft cylinder. 4. Cracked rockshaft cylinder cover.

If weighted lift arms fall too fast with drop/lock valve open, but not when valve is closed, check for: 1. Leaking or incorrectly adjusted lowering valve. 2. "O" rings between rockshaft control valve and rockshaft housing leaking. 3. Scored rockshaft valve spool or housing. 4. Stop valve leaking or damaged.

**141.** To check for proper operation of the steering priority valve, turn steering wheel and raise lift arms at the same time. Steering effort should be the same while lift arms are raising and after reaching the top. Disassemble and repair the priority and flow divider valves as outlined in paragraph 148 if difference is noticed.

**142.** To check volume of flow from hydraulic pump, detach outlet line from pump and attach a suitable flow tester. Start engine and operate hydraulic system until oil reaches normal operating temperature, then run engine at 3450 rpm and measure volume of flow. Correct volume is as follows.

655 Model . . . . . . . . . . . . . . . . . . 16 L/min. (4.2 gpm)
755, 756, 855 and 856 Models . . 22.5 L/min. (6 gpm)
955 Model . . . . . . . . . . . . . . . . . . 26.5 L/min. (7 gpm)

**143.** Check system relief valve pressure as follows: If equipped with selective control valve, attach pressure test gauge to an outlet port.

If not equipped with optional selective control valve, remove plug from test port on top of rockshaft inlet block and attach a suitable pressure gauge as shown in Fig. 89.

*Fig. 89—Hydraulic system relief pressure can be checked by attaching test gauge in place of plug in test port as shown.*

On all models, start engine and operate hydraulic system until oil reaches normal temperature, then run engine at about half throttle. On models without selective control valve, close rockshaft drop/lock valve completely. On models with selective control valve, pressurize the port. On all models, record the pressure indicated by gauge. Correct pressure for 655, 755, 756, 855 and 856 models is 13,652-14,617 kPa (1980-2120 psi) and for 955 models correct pressure is 16,665-17,650 kPa (2417-2560 psi). Pressure can be changed by adding or removing shims (3—Fig. 90) after removing cap (1).

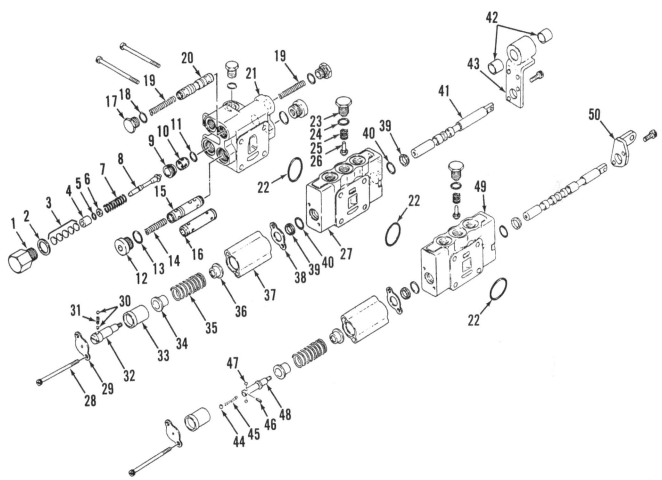

*Fig. 90—Exploded view of flow divider and selective control valves. The system relief valve (1-11) is attached to flow divider valve housing. Spacer (16) is used in place of (14 & 15) on models without selective control valves.*

1. Relief valve cap
2. Gasket
3. Shims
4. Bushing
5. "O" ring
6. Washer
7. Relief valve spring
8. Poppet
9. Valve seat retainer
10. Valve seat
11. "O" ring
12. Plug (two used)
13. "O" ring (two used)

14. Spring
15. Priority valve spool
16. Spacer
17. Plug (two used)
18. "O" ring (two used)
19. Springs (two used)
20. Proportional flow divider spool
21. Housing
22. "O" rings (three used)
23. Plug
24. "O" ring
25. Spring
26. Check valve poppet

27. Housing (front selective control valve)
28. Socket head screws
29. Plate
30. Detent balls
31. Spring
32. Detent screw (front valve)
33. Detent holder
34. Spring retainer
35. Centering spring
36. Spring retainer
37. Cover
38. Plate

39. Seal
40. "O" ring
41. Valve spool
42. Bushings
43. Support
44. Ball
45. Spring
46. Roll pin
47. Detent balls
48. Detent screw (rear valve)
49. Housing (rear selective control valve)
50. Support

## HYDRAULIC PUMP

### All Models

**144. R&R AND OVERHAUL.** To remove the hydraulic pump from rear of tractor, first remove the drawbar and drain fluid from system as outlined in paragraph 109. Loosen clamps, then move the suction tube (1 and 2—Fig. 91) out of the way. Unbolt and remove pto shield, disconnect outlet line from pump, then unbolt and remove hydraulic pump with suction tube. Withdraw and inspect pump drive coupling (91).

Unbolt and remove inlet fitting (3) from pump, then unbolt and remove pump cover (23 or 37). Remove seals, pump bushings, gears and shafts. Remove snap ring (12) from pump (B) of 955 models. On all models remove drive shaft seal (13). Make sure that all parts, especially the small fluid passages are clean, then inspect all parts for scoring, wear or damage.

Renew all seals and damaged parts. Coat all parts with hydraulic fluid when assembling. Press pump drive shaft seal (13) into bore until flush with housing surface of 655, 755, 756, 855 and 856 models or just below snap ring groove of 955 models. Install snap

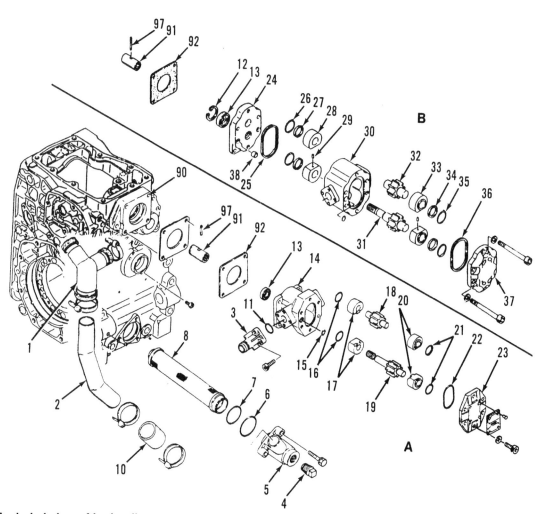

*Fig. 91—Exploded view of hydraulic pump and related parts typical of all models. Pump "B" is 26.5 L/min (7 gpm) used on 955 models; pump "A" is typical of all other models.*

| | | |
|---|---|---|
| 1. Suction tube boot | | 32. Pump gear |
| 2. Suction tube | 13. Seal | 33. Bearing block |
| 3. Pump inlet fitting | 14. Pump housing | 22. Seal ring | 34. Packing ring |
| 4. Rear drain plug | 15. "O" ring | 23. Pump cover | 35. "O" ring |
| 5. Adapter | 16. "O" rings | 24. Pump flange | 36. Seal |
| 6. Back-up ring | 17. Bearing blocks | 25. Seal | 37. Pump cover |
| 7. "O" ring | 18. Pump gear | 26. "O" ring | 38. Seal |
| 8. Filter screen | 19. Pump gear | 27. Packing ring | 90. Pump adapter |
| 10. Hose | & drive shaft | 28. Bearing block | 91. Coupling |
| 11. "O" ring | 20. Bearing blocks | 29. Pin | 92. Gasket |
| 12. Snap ring | 21. "O" rings | 30. Pump body | 97. Pin |
| | | 31. Pump gear & shaft | |

ring (12) on 955 models. Complete assembly making sure that all "O" rings and back-up rings are properly positioned in grooves, then install cover (23 or 37). Tighten cover retaining screws to 25 N.m (18 ft.-lbs.) torque. Tighten the four screws attaching pump inlet fitting (3) to 25 N.m (18 ft.-lbs.) torque.

When installing pump, position coupling (91) on pump shaft and attach suction pipe (1 and 2) to pump inlet fitting (3). Attach suction pipe to lower adapter (5) and hose (10), then tighten pump attaching screws and stud nuts to 25 N.m (18 ft.-lbs.) torque. Install pto shield and tighten retaining screws to 90 N.m (66 ft.-lbs.) torque. Refer to paragraph 109 for bleeding procedure.

## ROCKSHAFT AND HOUSING

### All Models

**145. REMOVE AND REINSTALL.** To remove rockshaft housing, first remove seat, fenders and any mounted implements. Remove cotter pin, then remove pin (A—Fig. 92) to detach rod from left side lift arm. Remove shoulder bolt (B) and detach pto selector from fender bracket. If equipped with front-wheel drive, detach rod from selector. Remove cotter pin (A—Fig. 93), then detach park brake rod (B) from lock lever. Remove shoulder bolt (C) and the two screws (D) to remove fender bracket (E). Unbolt and remove the seat bracket (F). Make sure that lift arms are lowered and hydraulic pressure is removed from system, then disconnect the two lines from flow divider valve and the four lines from selective control valves. Detach pressure line from hydraulic pump and remove the line. Remove the eight cap screws and one stud nut, then lift rockshaft from tractor.

Reinstall by reversing removal procedure. Tighten retaining screws and nut to 53 N.m (38 ft.-lbs.) torque. Use thread lock on shoulder bolt (B—Fig. 92).

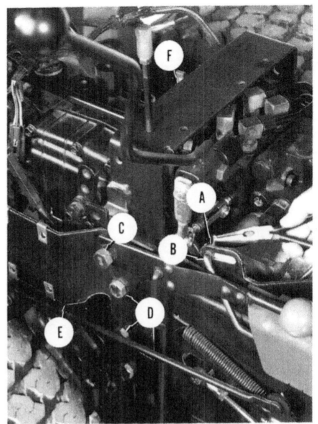

Fig. 93—View of tractor showing parts and screws to be removed from right side before removing rockshaft housing.

Check fluid level after assembling and operating system.

**146. OVERHAUL.** Refer to appropriate paragraphs for removal procedure and service to flow divider valve, selective control valves, piston cover and related parts. Detach feedback linkage (2—Fig. 94), remove feedback arm (3) from right side and retainer washer (20) from left side. Check ends of rockshaft (14) and lift arms (4 and 19) for alignment marks, which should be aligned when assembling. The marked spline of lift crank (10) should also be aligned with marked spline of rockshaft. Remove both lift arms (4 and 19), retainers (5 and 18), splined sleeves (7 and 16) and "O" rings (6 and 17). "O" rings (8 and 15) are in grooves on ends of splined sleeves (7 and 16). Slide rockshaft (14) from lift crank (10), then withdraw arm and piston rod (12) from housing. Piston (23) can be removed from top (front) of cylinder after removing cylinder cover (27).

Clearance between splined sleeve (7 or 16) and bushing (9 or 13) should not exceed 0.4 mm (0.016 in.). Outside diameter of sleeves (7 and 16) should be 44.95-44.975 mm (1.769-1.77 in.). Inside diameter of new bushings (9 and 13) should be 45.00-45.039 mm (1.771-1.773 in.) and should not require resizing after

Fig. 92—View of tractor showing location of pin (A) and shoulder bolt (B) which must be removed from left side before removing rockshaft housing.

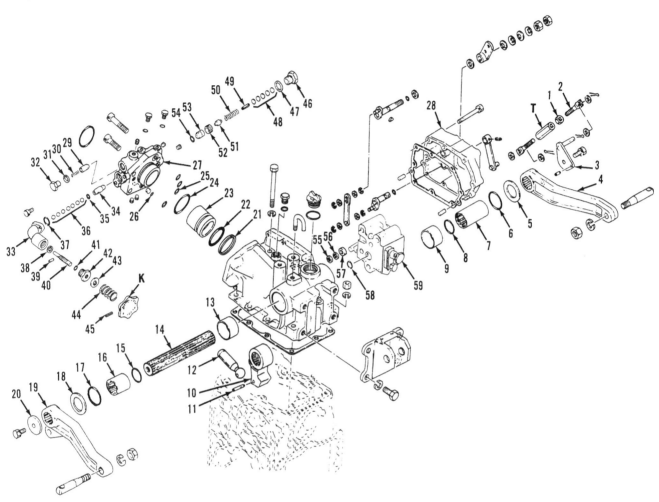

*Fig. 94—Exploded view of rockshaft and associated parts. Knob (K) is shown in Fig. 86 and turnbuckle (T) is shown in Fig. 88.*

1. Locknut
2. Linkage rod
3. Feedback arm
4. Right lift arm
5. Retainer (same as 18)
6. "O" ring (same as 17)
7. Splined sleeve (right side)
8. "O" ring (same as 15)
9. Bushing
10. Lift crank
11. Pin
12. Piston rod
13. Bushing
14. Rockshaft
15. "O" ring (same as 8)
16. Splined sleeve (left side)
17. "O" ring (same as 6)
18. Retainer (same as 5)
19. Left lift arm
20. Retainer washer
21. Back-up ring
22. Seal ring
23. Piston
24. "O" ring
25. "O" rings
26. "O" ring
27. Cylinder cover
28. Cover
29. Check relief valve
30. Spring
31. Gasket
32. Plug
33. Rockshaft drop/lock valve housing
34. Valve
35. "O" ring
36. Seven steel balls (8.7 mm)
37. Seal ring
38. Washer
39. Pin
40. Screw
41. Packing
42. Guide
43. Washer
44. Spring
45. Roll pin
46. Plug
47. Seal ring
48. Shims
49. Roll pin
50. Spring
51. Implement relief valve poppet
52. Seat retainer
53. Seat
54. "O" ring
55. Screen (2 used)
56. Seal (2 used)
57. Spacer (2 used)
58. "O" ring
59. Rockshaft control valve

installation. Outer edge of bushing should be 4.2 mm (0.166 in.) below outer surface of housing.

Assemble piston rod (12) to lift crank (10) and install retaining pin (11). Position lift crank and piston rod in housing and install rockshaft with marked splines of lift crank and rockshaft aligned. Stick "O" rings (8 and 15) in grooves of splined sleeves (7 and 16), grease bushing surface of sleeves, then slide sleeves into position with "O" rings toward center, against lift crank (10). Grease "O" rings (6 and 17), then install "O" rings followed by retainers (5 and 18). Install lift arms (4 and 19) with marked splines of arms aligned with marked splines of rockshaft. Install retainer washer (20) and feedback arm (3) tightening retaining screws to 52 N·m (38 ft.-lbs.) torque. Tighten screws retaining cover (28) diagonally to 27 N·m (20 ft.-lbs.) torque.

## FLOW DIVIDER AND SELECTIVE CONTROL VALVES

### All Models

**147. REMOVE AND REINSTALL.** Remove operator's seat and cotter pin (A—Fig. 93). Unbolt seat bracket (F) from fenders, loosen shoulder bolts (B—Fig. 92 and C—Fig. 93), then tilt seat bracket toward rear. Make sure that lift arms are lowered and hydraulic pressure is removed from system, then disconnect the two lines from flow divider valve and the four lines from the selective control valves. Remove the four socket head screws attaching the flow divider and selective control valves to cylinder cover and remove valves.

Stick sealing rings (22—Fig. 95) in position with grease, align flow divider valve and selective control valves, then insert the four socket head screws through valves before attaching to cylinder cover. The four screws will assist in holding parts in correct alignment while installing. Tighten retaining screws diagonally to 17 N·m (12 ft.-lbs.) torque and check valves for proper smooth operation. Uneven torque may cause sticking. Always use two wrenches to tighten hydraulic connections. Tighten union connectors to flow divider valves to 47 N·m (35 ft.-lbs.) torque and line nut to 34 N·m (25 ft.-lbs.) torque. When attaching lines to selective control valves, attach swivel, but leave swivel nut loose until after tightening lines to swivel. Tighten lines and swivel nut to 27 N·m (20 ft.-lbs.) torque. Complete assembly by reversing removal procedure.

**148. OVERHAUL FLOW DIVIDER VALVE.** Refer to paragraph 147 and remove flow divider valve. Refer to Fig. 95 and remove plugs (12 and 17) so that priority control valve and flow divider valve can be withdrawn from housing. Free length of spring (14)

should be 66 mm (2.6 in.) and spring should exert 54 N (12 lbs.) when compressed to 31.4 mm (1.24 in.). Springs (19) should have free length of 58.5 mm (2.3 in.) and should exert 4.9 N (1.1 lbs.) pressure when compressed to 46 mm (1.81 in.). Both springs (19) should have identical test specifications. Refer to paragraph 150 for service to system relief valve (1-11).

Lubricate all parts with hydraulic fluid when assembling. End of spool (15) with two holes should be toward right side of housing (21), which has port for hydraulic line from pump. Spacer (16) should be installed in place of spring (14) and spool (15) in models without selective control valves. Tighten Allen head plugs (12) to 98 N·m (72 ft.-lbs.) torque. Slotted end of spool (20) should also be installed toward right side of housing (21), which has inlet port. Tighten hex plugs (17) to 44 N·m (33 ft.-lbs.) torque.

**149. OVERHAUL SELECTIVE CONTROL VALVES.** Refer to paragraph 147 and remove the selective control valves. Front and rear valves are similar, but different as shown in Fig. 95. Keep parts from the two valves separate, even if parts were originally the same.

To disassemble the front selective valve, remove plug (23), then withdraw check valve (26) and spring (25). Detach linkage from valve spool (41), remove screws (28) and withdraw spool and related parts (29-40) from housing. Spool and housing are not available for service and complete new valve should be installed if either is damaged. Seal and "O" ring can be removed from other side of valve after unbolting and removing support (43). Be careful not to lose detent balls (30) and spring (31). Threads of detent screw (32) are coated with "Loctite" or equivalent. Coat spool and housing bore with hydraulic fluid while assembling. Coat centering and detent parts (30-36) with "Lubriplate" or equivalent. Tighten screws retaining support (43) to 11 N·m (96 in.-lbs.) torque

Disassembly and service procedures for rear control valve are similar to front valve, but it is important to keep parts of the two valves separate. Threads of detent screw (48) are coated with "Loctite" or equivalent. Ball (44) and spring (45) are retained by roll pin (46). Tighten screws retaining support (50) to 11 N·m (96 in.-lbs.) torque.

## RELIEF VALVES

### All Models

**150. SYSTEM RELIEF VALVE.** The system relief valve is held in place by cap (1—Fig. 95). System relief valve can be serviced separately or in conjunction with priority and flow diverter valves as outlined in paragraph 148. Parts (2 through 8) will be free when cap (1) is removed. Use a large screwdriver to remove seat retainer (9) so that new seat (10) can be installed. Free length of spring (7) should be 50.7 mm (2 in.) and spring should exert 325 N (73 lbs.) pressure when compressed to 42 mm (1.65 in.). Side of retainer (9), which has four slots, should be toward seat (10). Tighten retainer (9) to 29 N·m (22 ft.-lbs.) torque. System relief pressure is changed by varying thickness of shims (3). Refer to paragraph 143 for specifications and testing procedure.

**151. IMPLEMENT RELIEF VALVE.** The implement relief valve is located at (46-54—Fig. 94). To disassemble, remove plug (46), then remove implement relief valve and spring. Use a large screwdriver to remove seat retainer (52), then remove seat (53). Free length of spring (50) should be 44 mm (1.73 in.) and should exert 170 N (38.2 lbs.) when compressed to 38 mm (1.5 in.).

Coat all parts of valve in clean hydraulic oil before assembling. Be sure to install new "O" ring (54) if seat (53) was removed, then tighten seat retainer (52) to 27 N·m (20 ft.-lbs.) torque and plug (46) to 78 N·m (58 ft.-lbs.) torque.

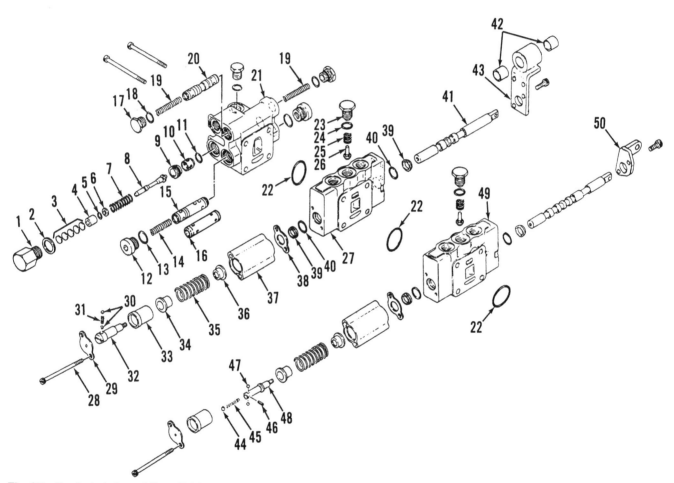

*Fig. 95—Exploded view of flow divider and selective control valves. The system relief valve (1-11) is attached to flow divider valve housing. Spacer (16) is used in place of (14 & 15) on models without selective control valves.*

| | | | |
|---|---|---|---|
| 1. Relief valve cap | 14. Spring | 27. Housing (front selective control valve) | 39. Seal |
| 2. Gasket | 15. Priority valve spool | | 40. "O" ring |
| 3. Shims | 16. Spacer | 28. Socket head screws | 41. Valve spool |
| 4. Bushing | 17. Plug (two used) | 29. Plate | 42. Bushings |
| 5. "O" ring | 18. "O" ring (two used) | 30. Detent balls | 43. Support |
| 6. Washer | 19. Springs (two used) | 31. Spring | 44. Ball |
| 7. Relief valve spring | 20. Proportional flow divider spool | 32. Detent screw (front valve) | 45. Spring |
| 8. Poppet | 21. Housing | 33. Detent holder | 46. Roll pin |
| 9. Valve seat retainer | 22. "O" rings (three used) | 34. Spring retainer | 47. Detent balls |
| 10. Valve seat | 23. Plug | 35. Centering spring | 48. Detent screw (rear valve) |
| 11. "O" ring | 24. "O" ring | 36. Spring retainer | |
| 12. Plug (two used) | 25. Spring | 37. Cover | 49. Housing (rear selective control valve) |
| 13. "O" ring (two used) | 26. Check valve poppet | 38. Plate | 50. Support |

# ROCKSHAFT CONTROL VALVE

## All Models

**152. R&R AND OVERHAUL.** Remove operator's seat and right fender. Unbolt seat bracket from left fender, then loosen shoulder bolt (B—Fig. 92). Unbolt and remove the rear seat brace from 655 models. On all models, remove shoulder bolt (A—Fig. 96), two bolts (B) and the fender bracket. Remove cotter pin and washer (C) and disconnect linkage (T). Tip seat bracket froward, then unbolt and remove cover (28—Fig. 94). Remove the three hex head metric screws attaching rockshaft control valve to side of rockshaft housing, then remove valve.

NOTE: Do not disturb locknut (5—Fig. 97) or lower adjusting stop screw (4). Setting is adjusted at the factory and may not be successfully accomplished as a service procedure.

*Fig. 96—Refer to text for removing the rockshaft control valve.*

Before disassembling valve, remove and clean the two 100 mesh screens (55—Fig. 94) from rockshaft housing ports, which align with valve ports. Bend locking tab away from screw (1—Fig. 97), then remove screw (1) and plate assembly (2, 3, 4 and 5). Withdraw spool and spring assembly (6, 7, 8 and 9) from housing (59). Remove plug (10) and load check valve (12 and 13). Remove plug (14), spring (15) and lowering valve (16, 17 and 18). Insert small Allen wrench or stiff wire into hole for spool (6) to push flow control spool (21) and related parts (19, 20, 22) from bore. Remove plug (23), "O" ring (24), spring (25) and unloading valve (26).

When reassembling, install new "O" rings and coat valves and bores with clean hydraulic fluid. Assemble in reverse of disassembly procedure. Clean threads of screw (1) and spool (6) and use "Loctite" or equivalent to lock threads when assembling. Make sure that cross pin of stop (9) is perpendicular to valve housing mounting surface when screw (1) is tight. Tighten screw (1) to 10 N·m (88 in.-lbs.) torque, then bend tab of lock plate (2) around flat of screw (1).

Install screens (55—Fig. 94) in bores of rockshaft housing with wide face of screen toward outside, then install seals (56) and spacers (57). Install "O" ring (58) in groove of valve housing (59), then attach valve to rockshaft housing. Tighten the three hex head metric screws attaching valve to 24 N·m (18 ft.-lbs.) torque. Install cover (28) using new gasket and tighten retaining screws evenly to 27 N·m (20 ft.-lbs.) torque. Tighten the two fender bracket screws (B—Fig. 96) to 149 N·m (110 ft.-lbs.) torque. Refer to paragraph 139 for adjusting turnbuckle (T).

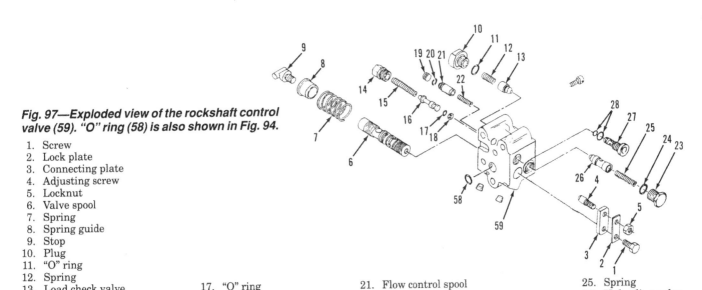

*Fig. 97—Exploded view of the rockshaft control valve (59). "O" ring (58) is also shown in Fig. 94.*

1. Screw
2. Lock plate
3. Connecting plate
4. Adjusting screw
5. Locknut
6. Valve spool
7. Spring
8. Spring guide
9. Stop
10. Plug
11. "O" ring
12. Spring
13. Load check valve
14. Plug
15. Spring
16. Lowering valve

17. "O" ring
18. Back-up ring
19. Plug
20. "O" ring

21. Flow control spool
22. Spring
23. Plug
24. "O" ring

25. Spring
26. Unloading valve
27. Valve seat
28. Seals

## CYLINDER COVER AND PISTON

### All Models

**153. R&R AND OVERHAUL.** Refer to paragraph 147 and remove flow divider and selective control valves. Remove the four socket head screws retaining cylinder cover (27—Fig. 94), then lift cover and rockshaft drop/lock valve. Remove plug (32) and withdraw spring (30) and check relief valve (29). Free length of spring (30) should be 27.7 mm (1.09 in.) and spring should exert 10.1 N (2.27 lbs.) when compressed to 18 mm (0.71 in.).

Unbolt and remove rockshaft drop/lock valve housing (33). Valve (34) can be withdrawn from housing bore using a M5 screw. Do not lose any of the 8.7 mm ($^{11}/_{32}$ in.) diameter balls (36). Free length of spring (44) is 36 mm (1.4 in.) and spring should exert 59.2 N (13.3 lbs.) pressure when compressed to 14 mm (0.55 in.).

Refer to paragraph 151 for service to implement relief valve assembly (46-54).

Piston (23) can be withdrawn from cylinder by threading a M6 screw into top of piston. Inside diameter of cylinder bore is 60.000-60.048 mm (2.362-2.364 in.) and outside diameter of piston is 59.94-59.97 mm (2.360-2.361 in.). Clearance between piston and cylinder should not exceed 0.3 mm (0.012 in.).

Install "O" ring (22) and back-up ring (21) in groove of piston, coat rings, piston and cylinder with hydraulic fluid, then insert piston in bore. Make sure that ends of back-up rings are not overlapped and that rings are not damaged while inserting into cylinder. Coat all parts of valves in clean hydraulic oil before assembling. Be sure to install new "O" ring (54) if seat (53) was removed, then tighten retainer (52) to 27 N·m (20 ft.-lbs.) torque and plug (46) to 78 N·m (58 ft.-lbs.) torque. Insert valve (34) with "O" ring (35) into bore of cylinder cover and grease the seven balls (36) before assembling. Tighten the socket head screws attaching housing (33) to cylinder cover to 25 N·m (18 ft.-lbs.) torque. Coat parts of check relief

valve (29-32) with clean hydraulic oil and assemble, tightening plug (32) to 34 N·m (25 ft.-lbs.) torque.

Before installing cylinder cover, install and lubricate new "O" ring (24), then stick "O" rings (25 and 26) in position with grease. Install cylinder cover and install the two longer retaining screws in top holes. Tighten the four retaining screws diagonally and evenly to 88 N·m (65 ft.-lbs.) torque. Refer to paragraph 147 to install flow divider and selective control valves. Refer to Fig. 98 for view of connecting linkage and operating lever.

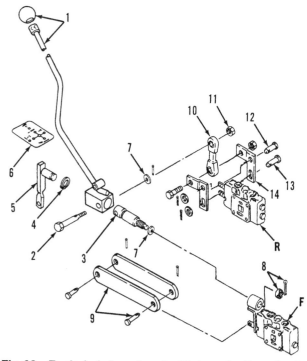

Fig. 98—Exploded view of control linkage for the selective control valves. Front control valve (F) and rear control valve (R) are shown exploded in Fig. 95.

| | |
|---|---|
| 1. Control lever | |
| 2. Bolt | 9. Linkage yokes, pins |
| 3. Pivot | & cotter pins |
| 4. Snap ring | 10. Ball joint |
| 5. Lever | 11. Nut (M6) |
| 6. Decal | 12. Pin |
| 7. Washer | 13. Pin |
| 8. Nut & cotter pin (M12) | 14. Lever |

# MAINTENANCE LOG

| Date | Miles | Type of Service |
|------|-------|-----------------|
|      |       |                 |
|      |       |                 |
|      |       |                 |
|      |       |                 |
|      |       |                 |
|      |       |                 |
|      |       |                 |
|      |       |                 |
|      |       |                 |
|      |       |                 |
|      |       |                 |
|      |       |                 |
|      |       |                 |
|      |       |                 |
|      |       |                 |
|      |       |                 |
|      |       |                 |
|      |       |                 |
|      |       |                 |
|      |       |                 |
|      |       |                 |
|      |       |                 |
|      |       |                 |
|      |       |                 |
|      |       |                 |
|      |       |                 |